Matrixial Breath

Day by day, the global climate crisis intensifies, yet there are little changes in the behavior of privileged societies. This book reveals that such paradoxical behavior is due to the prevalence of a narrative that considers planet Earth as a self-sustaining system, driven by a circular respiration pattern from one actor to the next. It argues that transforming this prevailing planetary breathing pattern is necessary to change humans' destructive behavior towards the more-than-human environment. The volume, in particular, references the feminine Matrixial Theory of psychoanalyst and philosopher Bracha L. Ettinger. In linking the Matrixial Theory to environmental issues for the first time, it explores the rhythms and scopes of a Matrixial breath and what alternative forms of relationship between humans and nature might emerge from it. It lays the foundation for an urgently needed subversion of thought and action toward novel ethics of breathing-with beyond homeostatic reductionism.

The first of its kind, this volume will be indispensable for students and researchers of environmental ethics, feminist thought, climate studies, social change, or critical theory. It will also be useful to cultural scientists, artists, philosophers, ecologists, theologians, architects, therapists, social workers, educators, and politicians.

Christoph Solstreif-Pirker, PhD, is Professor of Aesthetic Education at the University College of Teacher Education Styria (Austria), a practicing psychotherapist, and an interdisciplinary artist. His artistic and academic work focuses on performative research, encounter-investigations, painting, drawing, sound, and text. After studies at the University of Music and Performing Arts Graz, the Royal Danish Academy of Fine Arts in Copenhagen, and Graz University of Technology, he completed his PhD in artistic research and contemporary art, followed by an MA in philosophy and psychoanalysis under the supervision of Bracha L. Ettinger. His dissertation, titled "Being-Together-With the World-Without-Us," explored questions of space and subjectivity in the (Post-)Anthropocene, proposing feminine-performative modes of thinking and acting in response to contemporary planetary trauma. His work has been widely recognized and published in journals such as *PCS: Psychoanalysis, Culture & Society*, *Comparative Literature Studies*, *Ruukku: Studies in Artistic Research*, and *JAR: Journal for Artistic Research*.

Routledge Critical Perspectives on Breath and Breathing

Series Editors: **Lenart Škof**, Science and Research Centre Koper; **Magdalena Górska**, University of Utrecht

This book series focuses on a multidisciplinary range of perspectives on breath and breathing in a number of different philosophical, cultural, environmental and societal contexts. The series' main objective is to bring together ideas and offer a strong collection of original work centred on breathful life-worlds and breathable futures as the basis for a respiratory paradigm shift in the humanities and social sciences.

Based on a careful process of identifying and a detailed analysing of central respiratory issues (ie. breath, breathing, climate, and atmosphere), the series addresses the unevenly shared and felt affects and effects of human and non-human relationships to the air and to the processes of breathing. The books in the series will focus on breath, breathing, and air as a content, a vehicle and an atmospheric context for developing new theorizations of philosophical, ethical, cultural, social and environmental relations/ways of being-in-the-world. The interdisciplinary approach that combines perspectives of social sciences and humanities will address, for example, theologies, philosophies, geographies, medical perspectives, literary and political practices that are embodied in respiration. The books will address, but will not be limited to: phenomenological ontology of breathing, philosophical primacy of breathing, aerial poetics, police choke holds, police gassing; feminist, LGBTQ+ and intersectional politics; respiratory illnesses such as COVID-19 and black lung disease; wildfires, dust storms and other forms of unequal exposure to air pollution; climate change; theological traditions and belief systems about breath; and also more artistic renditions and interpretations of aerial relationships. By analysing and conceptualising these issues, books and authors in the series will contribute to cultivating new ways to understand the respiratory world and strategies towards a more just social relations and economy of the breath.

Phenomenological Ontology of Breathing
The Respiratory Primacy of Being
Petri Berndtson

Matrixial Breath

Aesthetically Respiring into the Trauma of the Present

Christoph Solstreif-Pirker

Routledge
Taylor & Francis Group
LONDON AND NEW YORK

First published 2026
by Routledge
4 Park Square, Milton Park, Abingdon, Oxon OX14 4RN

and by Routledge
605 Third Avenue, New York, NY 10158

Routledge is an imprint of the Taylor & Francis Group, an informa business

British Library Cataloguing-in-Publication Data
A catalogue record for this book is available from the British Library

ISBN: 978-1-032-99350-8 (hbk)
ISBN: 978-1-032-99351-5 (pbk)
ISBN: 978-1-003-60367-2 (ebk)

DOI: 10.4324/9781003603672

Typeset in Sabon
by Deanta Global Publishing Services, Chennai, India

To Goran, a great teacher

Contents

Introduction

This book is an attempt to become aware.

As I write these introductory lines, I am sitting at the window of a magnificent 19th-century mansion, looking out over the harbor of the small Greek island of Hydra, where I am staying for much of the work on this book. On the opposite side of the harbor basin, lined up steeply on a rocky hillside, I see small, elegant stone houses in shades of white and gray with tile-red clay roofs in the harsh light of the afternoon sun. None of these houses tries to stand out from their neighbors. In their simple elegance, they are different from one another and yet form a multifaceted urban relationship. The small village forms a specific structure in its fabric-like appearance and the almost classical design of the houses, between which narrow, canyon-like alleyways branch out like an endless maze. It is a place that inexplicably resonates with me, emanating a peculiar radiance evident even in the smallest details. It beckons to me from afar, compelling repeated visits, for it is here that I have discerned a subtle shift in my very breath, an altered state of being. This shift does neither ground in Hydra being a personal refuge from the academic tumult and its prominence as one of the most stunning places in the Aegean nor in the splendid hall-like chambers of my accommodation, the exquisite historical furnishings, the oil paintings, the Italian pointed arches, the sandstone color, or the cut glassware. In all these artifacts, spaces, and ambients, there is a simple, unpretentious, and vibrant embodiment of beauty, astutely identified by Roland Barthes in the paintings of Cy Twombly as "a kind of subtle energy which *makes it easier to breathe* (...) I float in the sky, I inhale the air" (Barthes 1991, 182). Through Barthes' lens, one comprehends how beauty and the aesthetic impression might engender an altered state of mind, a novel mode of perception, and respiration. Yet, upon a closer look, another determinant emerges. The aerated spaces of this house in which I find myself bear significance not merely for their precious superficiality, their aesthetic language, or their nostalgic patina, but rather, amid their historicity, they unveil a unique form of *melancholy*: The melancholy that such a reality, as displayed in the paintings, furnishings, art objects, ceilings, capitals, colors, sounds, and spaces of this interior, has *irrevocably passed away*. The privilege of working on my manuscript within one of the most exquisite buildings I have encountered so far does not conceal the fact that this building and its

DOI: 10.4324/9781003603672-1

encompassing radiance serve as a testament – a museum if you will – of a pre-modern ethos, a reality wherein people respired and existed differently. Amid its quasi-Arcadian ambiance, this domicile and its environs do not lull me into complacency; instead, they juxtapose a harsh and poignant contrast. It is no coincidence that Stefan Zweig's novel *The World of Yesterday: Memoires of a European* resonates profoundly in this setting. During my stay, I realized that this place's appeal does not stem from a veneer of apparent superficiality but rather from becoming aware of an *absence* – a void of a lost or obliterated time.

*

Such a perceived loss, as encountered in my home at Hydra, is paradigmatic for the loss of a way of being in the first quarter of the 21st century. It embodies the decline of the "subtle energy which makes it easier to breathe," the decline of a world nurturing vitality. Never before in human history has the world's population been so far removed from its original character as living human beings, with the ability "to find your way to the thing you feel when you love dearly, or when you create, or when you build your home, or when you give birth to your children or when you look at the stars at night" (Reich 1953, 175). Never before has the act of breathing – and thus, life itself – posed such a challenge for humanity; never before have individuals willingly and, at times, even pleasurably relinquished their humanity in reverence to the allure of a digital infrastructure that ostensibly tends to their every need; never before have people been as devoid of awareness concerning their fellow human beings, the earth, and divine transcendence; never before have people traversed the trajectory of their lives with such diminished aesthetic sensibility, such blunted perception; never before has a yearning for numbing anesthesia and painless oblivion been as pronounced as it is now; never before has the potential for interconnectedness among individuals been as vast as it is today – and yet paradoxically, never before have people been as profoundly disconnected from one another. These insights emerge amid the splendor of my accommodation, encapsulating the manifold losses incurred and the irreversible transitions witnessed. It is, in all likelihood, no mere coincidence that Leonard Cohen forged the following verses amid the environs of Hydra: "Your letters they all say that you're beside me now. / *Then why do I feel alone?* / I'm standing on a ledge and your fine spider web / is fastening my ankle to a stone." The traumatic immediacy of the present, where there is hardly any hold on the ledge, could thus be an endpoint and any further reflection rendered obsolete. However, it is precisely the notion of "despite" that constitutes the nucleus of this book. Despite teetering on the brink of the abyss, a pathway toward continued existence *persists*; despite humanity's plunge into barbarism, the potential for humane conduct *endures*; despite the brutalization of language, symbols of reconciliation *live on*; despite facing suffocating constraints, the capacity for respiration *survives*. A transformative encounter with Bracha L. Ettinger in November 2017, whose feminine "matrixial theory" serves as the book's cardinal point, imbues this "despite" with resolute justification. Ettinger's assertion that "after the catastrophe, the human must reappear" (Ettinger 2015b, 356) underscores the necessity of trauma for the unfolding of life. Paul Celan, whose oeuvre centers the notion of "despite," too, argues in a similar way: "language (...) had to pass through the thousand darknesses of deathbringing speech. It (...) passed through *could come to light again*" (Celan 2001, 395–396).

*

Drawing from these seminal references, the book explores how one might respire amid the contemporary planetary trauma. What respiratory procedures might be engendered when the air is stifled? How can one transcend hegemonic paradigms in biology, physics, psychology, and philosophy to attain a novel mode of respiration? What modes of relationship with fellow humans, the earth, and divine transcendence might be unveiled upon perceiving the trauma of the present, mourning the loss, and embracing the imperative of humanity? What constitutes "matrixial breath," and what significance does such a respiratory practice hold for humanity's relation with the cosmos and the need for environmental ethics? This book embarks on a speculative wandering into the terrains of these inquiries, unfolding and cultivating an awareness of their manifold ramifications. The book's formal cadence oscillates between firm and loose, between the clearly demarcated and the expansively conjectural. At times, thoughts are presented briefly and succinctly; at times, they require extended preparation. Moreover, concepts and references elucidating lines of thought are presented in concise interjections aimed at orienting rather than teaching the reader. Crucially, the book follows Barthes' dictum that "the reader is the space on which all the quotations that make up a writing are inscribed" (Barthes 1977, 148), signifying that the cogitations contained herein ought to be understood anew by the reader, reorganized and carried anew into the world; if this is the case, then the book has achieved its goal.

*

The book is divided into three parts. The first part ("Voluntary Suffocation") comprises an inventory of the present, delineating breathing patterns spanning distant epochs to the immediate past, invariably linked to their significance for human agency. This part culminates at the threshold of the present amid its desolate emptiness and capitulation to voluntary suffocation. Immediately succeeding is the second part ("Matrixial Breath – The Rhythms of Feel-Knowing Aerial"), which expounds upon the fundamental axioms of Ettingerian matrixial thinking. It elucidates the "despite" – the prospect of breathing within the void, within a vacuum – while delineating how such respiration diverges from established respiratory paradigms by its aesthetic character. The third part ("Metreorologic Speculations – Toward Ettingerian Environmental Ethics") constitutes the book's heart. Here, the central characteristics of "matrixial breath" are articulated as a *trans-inspiration* wherein the existence of the other, the transcendence of God, and the immanence of planetary matter float-into-each-other. This part culminates with conceiving matrixial aerials of love, engendering a novel scientific practice – *metreorology*. The conclusion revisits the first part, delineating potential avenues for instantiating matrixial breath within a socio-political framework transcending the anesthetizing power apparatuses of the present. It underscores the imperative of perceiving and confronting the present trauma, embracing the prevailing antipathy toward life affirmatively, to attain the horizons of a fragile yet mutually carrying relation with the world, always inherently inscribed within each human being.

Embracing melancholy is not depression but rather pertains to an encounter with beauty – a beauty of both void and fullness, as I am currently experiencing on the island of Hydra. Thus far, the discourse on such melancholy, beauty, and aesthetics has remained uncharted. This book serves as a pioneering exercise in raising awareness thereof – and in unfolding an intimate yet forgotten way of breathing.

Outline of Chapters

The first chapter critically examines the concept of the "Anthropocene," which has been proposed as a new geochronological epoch in recent years. The proposal of this epoch underscores the need to do justice to humanity's significant influence upon the planet, where human agency has permeated geological strata and ecological systems to such an extent that the dichotomy between "culture" and "nature" becomes untenable: a scientific assertion of human "creative will" in phenomena ranging from sediments to atmospheric particles. This capacity to transform the very essence of air assumes paramount significance not solely within the first chapter but across the different facets explored within the book. Although the proposal by the Anthropocene Working Group was formally rejected in March 2024, the Anthropocene continues to function as a vital conceptual framework. Interestingly, the scientific rejection goes together with the book's argument that the Anthropocene does not highlight a novel phenomenon exclusive to the last century; rather, it epitomizes the culmination of a historical trajectory intertwined with the emergence of medieval agricultural practices and the ethos of domesticating and reclaiming a seemingly "wild nature." These endeavors reveal an aspiration to efface the indeterminate "outside" in favor of a secured and delineated "inside." The threatening aspect of nature – what lies beyond the realm of human agency of the "I can" (Husserl 2020, 12) – is coercively internalized and colonized. With the assimilation not only of terrestrial realms but also of the aerial domain since the 17th century, the cut from the "outside" has been further advanced, where "atmosphere" literally means an encapsulated aerial sphere. In this context, reference to Deleuze's "grands milieux d'enfermement" (Deleuze 1992, 3) elucidates how the internalization of the outside parallels a profound transformation of the corporeal entities kept within this interior space. This inside space can be regulated externally, and the subjects can be fundamentally controlled. Drawing upon Foucault's conception of discipline and Lyotard's analysis of English society during Industrialization, it becomes apparent that such disciplinary mechanisms were not universally perceived as encroachments upon individual autonomy; instead, they were often construed as *desirable* impositions, for enclosure connotes security – a desired human aspiration. Nonetheless, the chapter refrains from exclusively attributing the enclosure practices of the Anthropocene to human agency universally. Instead, it underscores that the enactment of such internalizing narratives of war, destruction, and colonization primarily pertains to the practices of the white man. The chapter argues to more aptly characterize the Anthropocene as *White Androcene*, predicated upon subjugating a femininely connoted exteriority and suppressing a specific feminine agency within the planetary milieu. Such a feminine agency predominantly manifests as an aerial force – the effusive breath of "rûah," the "Spirit of God" – symbolizing an agency of receiving, carrying, trusting, touching, and the inclusivity of otherness and exteriority. The androcenic tendency for internalization, thus, has manifested a systematic eradication of this feminine spirit for centuries, culminating in the suffocation of all breathable air amid the atrocities of the Shoah: a total internalization.

The second chapter departs from the point in time that poet Paul Celan named a "breathturn": After the Second World War, the end of enclosure milieus can be discerned around the year 1950, when androcenic practices reached their climax and conclusion through the atomic bomb. This endpoint, however, does not mean the end of androcenic planetary transformation; instead, the established mechanisms of exclusion are changing. The overt power of discipline is giving way to a more covert power of control. In the control regimes described in detail by Foucault, there is no longer a hierarchical order nor a singular sovereign subjugating a broad mass to be excluded; rather, everyone has become an observer. This flattening and de-centering of discipline is also linked to a change in its expression: it no longer operates in a gloomy and despotic form but in a positive, collaborative style. Control no longer takes place "from above" but acts "together"; it no longer manifests authoritarian command but a friendly, quasi-inviting question: "We've seen you: why don't you join in?" Synchronization is a new disciplinary measure framed in idioms of bliss and an enforced sense of togetherness. Individuals become numbers, "dividuals" that can be shaped and replaced at will. Regimes of control are as smooth, soft, and frictionless as possible and characterized by the feigned absence of pain and suffering. In it, any form of weakness, compassion, and attention to the needs of the other is absent. This no longer requires cutting off air but the establishment of a common, synchronous breathing pattern to which all subjects must submit. Such a reality moves from the glass dome of discipline to a flat, monofunctional surface on which everyone floats side by side, controlling each other. Contrary to their soft, quasi-feminine expression, such spaces and practices are consistently organized phallocratically and based on rivalry and speed that do not allow for subversive movement. Regimes of control are evident in the flatness of everyday digital displays that profoundly organize human life and transform the breathing and living rhythms of the human subject. Breathing no longer takes place in an internalized space cut off from the outside but in a planetary flatness transformed into a "cover": no discipline, but synchronization. Human action becomes limited to its dependence on the uniform flatness, negating (and not only preventing) any reference to an unknown otherness. This flatness heralds the illusion that smoothness can be equated with invulnerability and that the pursuit of happiness is found in the constraints of togetherness and positivity. A more profound observation, however, shows that the resilient human is a human fundamentally deprived of humanity. The androcenic present seems to be a dimension of blissful barbarism – conceptually, and in all its worldly unfoldings.

The third chapter continues the preceding observations and diagnoses a fundamental lack of participation and perception within the synchronized regimes of control, which can also be characterized as "apathy towards reality" (Han 2021, 32). The smoother, happier, and more synchronized reality becomes, the more numb and disinterested humans become in their thoughts and actions. Such apathy is apparent not only toward fellow human beings and the world around us but also toward oneself: humans withdraw from themselves and give up their independence and ability to breathe. Out of a heightened need for security, they voluntarily surrender their remaining respiratory power to a digital breathing infrastructure. The androcenic human is equipped with an invisible breathing prosthesis that expresses itself in all forms of everyday digitality and fundamentally transforms the human body-psyche: Resistance, the sensation of pain and opposition are made impossible by it, inducing multifarious forms of rigidity and anesthesia. Particularly impressive in such apathy toward reality is the loss of any altruistic ability and the atrophy of the lung as an ethical organ that allows one to participate in others through breathing, absorb others into oneself, carry them, or suffer with them. Interestingly, even if narratives of co-existence are particularly present in the androcenic reality, the perception and absorption of the other into one's corporeality entail something fundamentally different than romantic narratives of living together in harmony. The current apathy toward reality concludes that a fundamental human characteristic – opening to and welcoming the other – is about to be abandoned. The human being now only fulfills the function of an interface to which the tube of a universal breathing machine is attached: the Androcene thus manifests as an epoch of intubation, coupled with a fundamental indifference to any otherness

– to other humans, other living beings, other thoughts, other worlds, other spaces. Because of its perfidy, Lyotard calls such indifference "the worst form of violence" (Lyotard 1993, 74), worse than any form of discipline or control. The voluntary abandonment of humanity and the unconditional surrender of perception can, in fact, hardly be surpassed in violence; the human becomes static, rigid, hopeless, and faithless.[1] At this point of outlining the present as the greatest possible catastrophe, the chapter reaches its analytical bottom. But precisely within the "valley of the shadow of death," a "despite" appears – a set of silent, inconspicuous, yet unavoidable questions: does the impending apocalypse also reveal a second, different path? Can breathing be actualized in a new form precisely by *embracing* the traumatic present? Does the affirmation of the prevailing darkness allow us to breathe unknown air with yet unknown procedures, rhythms, and processes? Is there a hidden vitality lying behind apathy? Can air be understood and conceived differently than with the mechanisms of apathy and anesthesia? Is the encounter with a vacuum perhaps even life-giving? With reference to François Laruelle's non-philosophy, this chapter opens up a different perspective that lies before any attribution, comprehensibility, and accessibility; the air becomes emptied and expanded. In the resonances of the contemporary requiem lies another aerial dimension that has so far remained unrecognized: if humans remove their breathing tubes, they will not suffocate.

The fourth chapter further speculates on the possibilities of breathing in a vacuum. With reference to the positions of philosopher E.M. Cioran, the chapter claims that emptiness is not negativity and that a fundamental, previously unrecognized potentiality is inherent in the voided, deserted, and wasted. A counter-position to the anxiety-driven emptiness of Heidegger's phenomenology is also outlined: emptiness unfolds a new vitality through which humans might return to the origin of their humanity. Entering the void means a re-encounter with an unfamiliar silence and the feminine agency of rûah. The first verse from the Book of Lamentations, "How doth the city sit solitary, That was full of people!" (Lam. 1:1), is directly linked to such silence, wherein the blowing of a faint yet mighty wind can be felt. The chapter illustrates how, after the phallocratic regimes of internalization, anesthesia, and intubation, a solitary reality, scorched earth, remains; technologies have collapsed, and synchronous atmospheres have shattered. Yet, despite this darkened wasteland, a light remains. Fragments from the poetic work of Paul Celan serve as support for its unheard-of thought: to see the present trauma, the time before the fall, as a *potentiality*. Right now, despite all this, or precisely because of all this, a possibility emerges to perceive something very close to heaven, life, and humanity; perhaps the androcenic apocalypse even allows breathing for the first time. At this point of understanding trauma in a non-indifferent way, the chapter introduces the feminine thoughts of the Israeli psychoanalyst, philosopher, and artist Bracha L. Ettinger, whose work will guide the rest of the book. Ettinger's feminine-maternal theory, which has become known as "matrixial theory" since the 1980s, illustrates a view of human subjectivity that, following Paul Celan, must first wait for catastrophe to rediscover humanity and even necessitates engagement with the traumatic, painful, dark and destructive to make the human(e) reappear. The course of this chapter introduces some of the main concepts of matrixial theory, which are elaborated further in the following chapters. The idea of "matrixial love" is fundamental, as it suggests a relational theory opposite to the romanticism of androcenic bliss. Matrixial love theorizes intimate relationships based not on economy or mechanical symbiosis but on fragile, asymmetrical procedures of trust, resonance, carrying/being carried, concern, and compassion. In matrixial love, the other is never a stranger but holds traces of my self within it: through her/his/its existence as a (non-)I, it gives the human a primordial assumption of responsibility for the other, together with the obligation to perceive and absorb this other in its uniqueness, its desires, and burdens. The other affects me because I am involved with and inscribed in her/him/it. With this Ettingerian perspective, breathing expands from a purely biological function into an ethical-aesthetic process. Vitality begins to emerge.

The fifth chapter further elaborates on the responsibility of being human from an Ettingerian perspective. In reference to Hebrew etymology, Bracha Ettinger shows that the "subject" must always be understood as a "carrier" – both literal and figurative: shouldering, carrying, caring. Hinting at the Cartesian dictum, Ettinger states that "I care-carry ergo sum" (Kaiser and Thiele 2018, 107). Such a responsibility is already evident in the opening chapters of the Hebrew Bible, where carrying is the first commandment to humans by God; however, it is often lost in translations: "If you will be doing what is good: carry" (113). Ettinger subsumes this commandment in the term "carriance," which means a non-fusional asymmetrical relationship based on sensing the echoes of the other: a relationship that understands itself as listening and responding beyond language, in the realm of fragile resonances, allusions, facial expressions, glances, or moods. Such carriance is not exclusively related to the subject's immediate other; it also refers to the other's past and future. Carriance carries beyond the obviously visible and symbolizable; it also includes times and spaces that lie beyond the present but are just as relevant. Ettinger shows that a path to humanity can only be identified if it points in a carrying direction – when the human being continuously decides to tread this path, to choose her/his original designation as a subject-carrier. This primordial attention to the other and the unconditional carrying of the other also goes hand in hand with a disempowerment of one's supremacy: the subject becomes fragile and permeable. It opens itself to the agency of the rûah – to the feminine resonances of the other drifting toward me and asking to be carried. Such a decision to break open the hegemony of the individual has hardly been made in Western cultural history and the androcenic present but has primarily been prevented, if necessary, with brute force. In the Ettingerian matrixial theory, such a perspective is now developed further, to the point where the I can only be constituted through the other, a position that also comes very close to Emmanuel Levinas' ethics. Breathing in such a matrixial condition is no longer a synchronous breath in which the subject becomes numb due to its desire for security, but, on the contrary, a carrying breath that only creates security through the realization of true humanity – through the decision to fragilize subjectivity and to transform it into a resonating trans-subjectivity. In this carrying breath, a co-poiesis emerges, full of "compassionate hospitality and generosity" (Ettinger 2005, 707). In contrast to prevailing androcenic conceptualizations such as autopoiesis and sympoiesis, Ettinger presents co-poiesis as a creation beyond fusion and sacrifice, in which the most diverse alterities, different Is and (non-)Is are interwoven. Identity is thus primarily constituted by the presence and reception of the other (understood as human, trans-human, more-than-human, or unknown), such that subjectivity may be composed (and constantly recomposed) much more comprehensively than previously assumed.

In the sixth chapter, the practices of matrixial becoming-permeable reveal that perceiving and receiving the other is a thoroughly aesthetic procedure. Here, the term *aisthesis* is understood in its original meaning as the sensual perception of the other, together with the openness to welcome her/him/it and to be affected by her/him/it. This aesthetic component of feminine-matrixial carriance is examined in contrast to the Heideggerian "dwelling" and the relational constellation of the "fourfold." It becomes apparent that both Heidegger and Ettinger situate "humanising the human" (Ettinger 2019, 183) within the aesthetic approaches of poetry and art. For Heidegger, however, such *aisthesis* is always bound to the concept of an "open center," which arises from the dialectical struggle between "clearing and concealing" (Heidegger 2001b, 178) and always remains inaccessible in its substance. Furthermore, for Heidegger, art always entails the "bringing-forth" of something new, created from an inaccessible nothingness. This Heideggerian view of aesthetics is closely related to sublime beauty, which remains alien in its grandeur; auto- or sympoietic systems are also characterized by similar inaccessibility, which walks hand in hand with the indifference and anesthesia of the androcenic present. As a counter-position to such an inaccessible aisthesis, Ettinger presents a matrixial aesthetic paradigm that consists of co-poietic carriance and a fundamental relationality. It is not about the "bringing-forth" of something new but about perceiving and being affected by what is there – about an awareness of the other's presence and the openness to be touched by the pervading rûah. Ettinger's aesthetic theory does not ground on the sublimity of the other; instead, the perception (aisthesis) of the wounds of the other and the subtle traces of the (non-)Is make the other related to me. Thus, the task of carrying, through which every human is determined, means the creative weaving of a shared fabric made possible by listening and opening up to each other. In carrying breath, a "co-inhabit(u)ation" (Kaiser and Thiele 2018, 122) takes place, which can pave the way for true humanization of the human. Breathing, understood aesthetically, fundamentally counteracts androcenic anesthesia and does not allow world-anxiety but *world-entry*

– the unfolding of a different aerial space of mutual *floating-into-each-other*. In this chapter, the aesthetic processes of floating-into-each-other are highlighted with reference to Ettinger's painterly practice wherein "subreal elements and strings resonate with-in and in-between human beings humanely in touch with the cosmos" (Ettinger 2019, 194). When the other is no longer recognized as a stranger but embraced as part of the self, the human subject no longer breathes for itself but for and with the other: a way out of the androcenic apocalypse is mapped out.

The seventh chapter investigates the question of actually realizing matrixial aesthetic breath. The chapter shows that – with reference to the etymological roots in the Hebrew Bible – human subjectivity is imbued with a virtual ability to float-into-each-other from the moment of its conception. Whether willingly or not, every human carries an aesthetic awareness of the other drifting toward him/her, an awareness that in the emptiness of space, in the wasteland of catastrophe, a wind blows that touches humans at the heart of their existence. Such actualization corresponds with perceiving the lung as an ethical organ, first proposed by Levinas, and the primordial awareness of a creative blowing wind at the origin of being. Building on Deleuze's and Guattari's concepts of territorialization and de-territorialization, the chapter shows, on the one hand, the importance of establishing one's territory (without which the human could not live), constituted of repetitions and rhythms similar to the periodicity of the beating heart or Levinas' notion of "enjoyment" (Cf. Levinas 1969). On the other hand, the necessity of the mutation of this very territory is emphasized, and the imperative to repeatedly set aside one's enjoyment in favor of the other: such a de-territorialization allows the (re-)experience of the feminine rûah in one's territory of identity and bodily integrity. The chapter asserts that without de-territorialization and receiving the other, humans cannot continue to stay alive: exclusively stuck in a safe territory impoverishes and degrades them to living dead. The biologically exclusive practice of re-spiration (the constant territorializing repetition of the same breathing pattern) is matrixially expanded into trans-spiration, a mutually perceptive and carrying breathing beyond symbiosis and blissful relationality. Referring to the theology of substitution and the concept of the Levinasian "face," trans-spiration is understood as restless breath, fundamentally opposed to the synchronous dynamics of the anesthetized present. At the same time, matrixial theory shows that such restlessness does not act with phallocratic power but rather represents feminine fragility. Trans-spiration takes place in a floating form, beyond vertical and horizontal orientations; it exhibits countless levels of gradations, just as in the "processes of birthing, disbirthing, unbirthing, a-birthing and birth" (Ettinger 2019, 184). In this floating orientation of subjectivity, the chapter introduces the notion of a feminine "*matrixial breath*" that attempts constant de-territorializations of its re-spiratory patterns and redefines subjectivity by welcoming the particles of the other – human, more-than-human, transcendent, and unknown. In matrixial breath, the former separation between humans and their environment has been definitively shown to be an indoctrination of androcenic power.

The eighth chapter further clarifies the implications of matrixial breath. It demonstrates that in addition to perceiving the other, trans-spiration always includes the experience of "total otherness," the goodness and love of God himself. It is precisely this "Spirit of God" that the androcenic present seeks to conceal with mechanisms of anesthesia and incubation. Aesthetically absorbing the breath of the other reveals a manifestation of the infinite, a sensing of the eternal breath. Trans-spiration thus becomes apparent as a biological-ethical-transcendental event. In the Psalms, the characteristic of this eternal breath is revealed as a "small voice" – the sound of flapping wings providing protection and shelter. This eternal breath carries the human "in a spacious place," offering protection and freedom. The chapter contrasts this eternal breath of God, which can be experienced in trans-spiration, with the Kantian *noumenon*. It is shown that God is not exclusively inaccessible and distant but that His power and breath can be experienced: God is far from any totality. The chapter traces a philosophical development that has established this inaccessibility precisely as a prominent component of androcenic anesthesia: in Kant, Heidegger's phenomenology, Freud's psychoanalysis, and the recent Object-Oriented Ontology, the world always manifests itself through instances of inaccessibility, appearing as threatening "hyperobjects" (Cf. Morton 2013). Contrary to this proclaimed inaccessible otherness, trans-spiration suggests a fundamental possibility for participating in this otherness, in the fellow human being, in the greatness of God, and in the "spacious place" of the world. Matrixial breath calls for a tripartite relationship to the world that situates the human as a participant in an interpersonal existence (trans-spiration), a divine transcendence (in-spiration), and a planetary immanence (trans-in-spiration). Being-in-the-world is no longer defined in terms of anxiety, as in Heidegger, but is permeated by a loving connection with the feminine rûah, the infinite breath of God. Becoming human(e) happens without threat and sacrifice, but rather, as Ettinger suggests, with caring awe, with a "concerned arousal" (Ettinger 2011, 3–4). It is never guided by fate but rather a gift that reveals itself to humans, which they are free to accept (de-territorialization) or reject (territorialization). Beyond the decision, however, lies the human foundation in the original breath of God, from which humans cannot separate, even amid the doctrines of anesthesia. In the depths of human beings, there always remains a possibility of opening up and encountering God's infinite and freedom-giving agency.

The ninth chapter finally returns to the air as the medium of trans-in-spiration in a tripartite world relationship. Here, the air is not understood as an atmosphere, a closed sphere of air, but rooted in the etymological context of the Proto-Indo-European root *lewb^{h}- (to desire, covet, want, to admire, praise, to love). Air becomes a medium of desire and love, allowing love to unfold as the essential condition of life, as Max Scheler suggests. Air is not understood as an ocean of air in which separate entities move independently of each other or unite with each other according to a biological law. Air in the matrixial understanding goes beyond the Aristotelian scientific perspective since established disciplines of ecology or meteorology can no longer serve to describe the aesthetic trans-in-spirative processes in the air. To explain the loving aerial condensations, the new science of *metreorology* is introduced. Metreorology is a feminine-matrixial extension of meteorology, located beyond anxiety-driven, mechanical, and positivist epistemes. Metreorology distances itself from androcenic discourses of "entanglement" by suggesting a science beyond determinable and functional usability, in which the "small voice" of the other can be perceived and received. Metreorology grounds in aesthetics and reveals the manifestations of matrixial eros, which unsettles the territory of subjectivity and opens up a dimension of silence in which the wafting of the other and the sigh of the total other of God can be experienced. However, metreorology is not only concerned with the processes of respiratory de-territorialization but it also addresses the question that every floating-into-each-other is only momentary and must end to start anew. From the freedom of trans-in-spiration, humans must always return to the anxiety of the prevailing anesthesia. The human is never entirely and forever displaced, but always returning: floating-into-each-other, carrying and being carried by the other, then torn apart again, set back in the original place. The hopeful component of metreorology lies precisely in this return: humans are always invited to float-into-each-other and to form a new identity; at any moment, they can breathe and allow themselves to be displaced into a space beyond anesthesia – provided they accept this gift given to them. The chapter concludes with the observation that humans are dust in every moment of their existence – ready to absorb and hold each other. Metreorology thus becomes a discipline of a new understanding of the relationship between humans, their fellow human beings, divine transcendence, and the earth. Its floating and feminine-matrixial paradigm implies a different view of planetary relatedness – turning the present apocalypse into a cradle of a different vitality and humanity.

Hydra, March 2024

Note

1 Cf. the monologue of *Stalker* in Andrei Tarkovsky's 1979 film of the same name: "If they only imagined how tired I am! Only God knows! And they call themselves intelligentsia. Those writers! Scientists! (…) They do not believe in anything. That their … organ with which they believe has atrophied! (…) My God, what kind of people they are …" (Tarkovsky 2002).

Part I

Voluntary Suffocation

1 Escaping the Outside

On 11 July 2023, while writing the first lines of this book, the "Anthropocene Working Group" (AWG) issued a press release announcing an agreement regarding the site to validate the "Anthropocene" hypothesis.[1] To adequately understand the relevance of this statement, we must look back a few years. In 2000, the Dutch meteorologist and atmospheric scientist Paul J. Crutzen, together with the US biologist Eugene F. Stoermer, claimed that the geochronological positioning of the present reality in the Holocene, which had lasted around 12,000 years, was no longer feasible. Human impact on the planet has accelerated to the point that the consequences of human practices can be traced in the geological and atmospheric layers of the earth. It is no longer possible to argue that the environment of the early 21st century is geologically identical to that of 10,000 BC. Crutzen and Stoermer asserted that due to the

> major and still growing impacts of human activities on earth and atmosphere, and at all, including global, scales, it seems to us more than appropriate to emphasize the central role of mankind in geology and ecology by proposing to use the term "Anthropocene" for the current geological epoch.
>
> (Crutzen & Stoermer 2000, 17)

With this statement, Crutzen and Stoermer claimed that transitioning from the 20th to the 21st century is nothing short of the entry into a new geological epoch – which did not start under "natural" conditions but "a-naturally" – manifested by phenomena such as the growth of the world's population, increasing urbanization, the depletion of fossil fuels, the high release of carbon dioxide, sulfur dioxide, nitrogen monoxide, and methane into the atmosphere, the colonizing transformation of the earth's surface, the increasing destruction of biodiversity, the enlargement of the ozone hole in the Antarctic, the pollution, overfishing, and acidification of the oceans, and, finally, the irreversible alteration of natural cycles.

*

DOI: 10.4324/9781003603672-3

Crutzen and Stoermer's suggestion has equally captivated experts in the natural sciences, humanities, and arts. In 2009, the Anthropocene Working Group was established as a worldwide scientific consortium to investigate and advise on Anthropocene-related issues, including three basic questions: What year or decade marks the beginning of the Anthropocene? (Related to this: In which social structure did the Anthropocene unfold?) Which compounds serve as obvious indicators for the Anthropocene? (Related to this: What human practices define the Anthropocene?) Finally, is there a site that can verify the Anthropocene hypothesis? (Related to: Is the Anthropocene a local or global phenomenon?) The answer to the last point allows conclusions to be drawn about the first two questions. Once a site is discovered that can best testify to the necessity of a geochronological repositioning, answers can be found about the constituent practices and the beginning of the Anthropocene itself. The AWG presented the preliminary results of its site search in July 2023. Crawford Lake in Ontario, Canada (43.4685761°, –79.9489985°) was chosen. Compared to its size, the lake has a significant depth (24 meters), prohibiting mixing of deeper and higher layers of water. The lake's isolated bottom layer allows for the creation of dark sediment layers (known as "varves"), which descend from the air to the bottom of the lake and contain information about the environment at the time of their sinking. Furthermore, light-colored calcium and carbonate ions are dissolved from the lake's rocks in the warmer surface water every summer, covering the sediment layers in patterns akin to tree rings. The analysis of sediment layer samples allows for a high-resolution annual representation of human practices over the past centuries (Cf. Anthropocene Working Group 2022). For example, maize and sunflower pollen from the late 13th century onwards attest to the indigenous people's agricultural practices near the lake, lasting for 200 years. The sediment layers containing charcoal particles date back to the early 19th century and reflect industrial activity (due to a nearby sawmill), along with increased nutrient input, erosion, and over-fertilization. In the early 20th century, carbonaceous particles and nitrogen isotopes from fossil fuels increased dramatically in the varves. Until this point, the remnants of human behavior in the sediment layers are mainly from local practices around the lake. However, this changes within the sediment layers from the 1940s and onwards, in which radionuclides such as Plutonium-239 and Caesium-137, as well as an increase in Carbon-14, can be detected, correlating with the atomic bombs dropped in 1945 and the US and Soviet nuclear weapons tests in the 1940s and 1950s. These findings demonstrate that the lake's sediment layers reflect not just human practices of the immediate local environment but also global actions that extend beyond Canadian territory, such that "Crawford Lake, Ontario, Canada is proposed for the *Global boundary Stratotype Section Point* (GSSP)." The analysis of its sediment layers also suggests that "the proposed GSSP level is dated by varve chronology to 1950" (McCarthy et al. 2023, 171). Crawford Lake allows conclusions about the Anthropocene to be drawn because its sediment layers provide both detailed evidence of local human practices from previous centuries, and actions unrelated to the lake's local context dated around 1950, such as nuclear weapons tests in Hawaii or the former USSR. The presence of radioactive remnants in Crawford Lake indicates that human activity has highly transformed the geological strata all over the planet: what was previously understood and perceived as "nature" or "natural" no longer exists. The Anthropocene hypothesis shows that not only can regional patterns of human behavior be found in the geological layers of the earth but that no site on the planet has remained untouched by human activity due to the emergence of nuclear weapons: "untouched wilderness" has ceased to exist.

*

The remnants emphasizing the need for a new geochronological epoch are not neutral but rather represent a specific ideology of war and destruction. Following this observation, the question arises as to whether this *violent* transformation of the planetary environment is a unique phenomenon of the 20th century, i.e., whether the Anthropocene is solely embedded in human practices of the 1940s and 1950s as assumed by the AWG or whether the Anthropocene rather marks the *point of culmination* of a cultural tradition that dates back several centuries. Such a view corresponds to the reasons given for rejecting the formal dating of the Anthropocene to the mid-20th century and the official definition associated with it, as Jan A. Piotrowski, a member of the AWG, has also noted: "It constrains, it confines, it narrows down the whole importance of the Anthropocene (...) What was going on during the onset of agriculture? How about the Industrial Revolution? How about the colonizing of the Americas, of Australia?" (Zhong 2024). Mike Walker, an earth scientist and professor emeritus at the University of Wales Trinity Saint David, added: "Human impact goes much deeper into geological time (...) If we ignore that, we are ignoring the true impact, the real impact, that humans have on our planet" (Zhong 2024). The fact that geologically transformative human behavior only became geologically visible and verifiable in the middle of the 20th century does not imply that these intentions did not exist much earlier and had grown in scope and dynamics – the "great acceleration" of the 20th century would thus have had its engine started much earlier. Lynn White Jr. points out that the transformation of the earth is tied to the creation of an agricultural tool already in use in the 7th century:

> By the latter part of the 7th century after Christ, however, following obscure beginnings, certain northern peasants were using an entirely new kind of plow, equipped with a vertical knife to cut the line of the furrow, a horizontal share to slice under the sod, and a moldboard to turn it over (...) It attacked the land with such violence that cross-plowing was not needed, and fields tended to be shaped in long strips (...) distribution of land was no longer on the needs of a family but,

> rather, on the capacity of a power machine to till the earth. Man's relation to the soil was profoundly changed. Formerly man had been part of nature; now he was the exploiter of nature.
>
> (White 1967, 1205)

Lynn's image of the medieval implementation of the "heavy plow" and the statement "man's relation to the soil was profoundly changed" present a shift from participative to exploitative and aggressive agricultural practices. The fact that the remnants of such violent action can be found not only in the furrows of medieval cultivated fields but also as radioactive fallout in the earth's layers simply indicates an intensification of human destructive power: a clear line of development can be traced between the invention of the "heavy plow" and the atomic bomb. This suggests that the Anthropocene's origins are considerably earlier than the 1950s, maybe as early as the 7th century. This historical repositioning of the Anthropocene would allow a compromise to be found with regard to the formal rejection of the Anthropocene hypothesis in March 2024 and the need to stay with the concept as an object of investigation – perhaps even understand the Anthropocene in its full depth for the first time.

*

In his examination of White's paper, Graessle (2018) suggests that

> before the introduction of the heavy plow into Western Europe, settlers were confined to light soil regions, since the antiquated ard could not clear areas that had dense soil or were forested. Once the heavy plow was introduced, however, it made small work of these areas and opened them up for cultivation and settlement.
>
> (30)

The fact that more aggressive conduct toward the environment allows for cultivation and settlement implies the disappearance or rather intentional elimination of what was previously "outside" the human realm. With the heavy plow, human activity has transformed a formerly inaccessible and untamed "outside" into a predictable and intelligible "inside." Such practices spread from the limited domain of the covered shed, the enclosed garden, to the nearby landscapes and territories until the entire world is "shaped in long strips" by the beginning of the 20th century. According to Peter Sloterdijk (2009a), the global phenomenon of making-the-outside-disappear is a "mystery of insulation/island-making [Insulierungsgeheimnis]," without which human evolution up to its dramatic climax in the 1950s remains elusive:

> Human evolution can only be understood if we also bear in mind the mystery of insulation/island-making [Insulierungsgeheimnis] that so

> defines the emergence of humans: Humans are pets that have domesticated themselves in the incubators of early cultures. All the generations before us were aware that you never camp outside in nature.
> (1)

Based on these fundamental traits, which Sloterdijk identifies even earlier than White – in fact, at the first appearance of Homo erectus some two million years ago – the Anthropocene may be viewed not as a mere geological epoch – just as the AWG had planned – but, much bigger and even more frightening, as a *trajectory of enclosure*. Human behavior, according to both historical and current geological findings, is inextricably linked to this desire for internalization, coupled with a violent dynamic to enforce and establish such internalizing procedures. The Anthropocene's claim that the planetary "naturalness" has ceased to exist is based on this internalizing tendency – the global colonization of unfamiliarity and exteriority. This tendency toward internalization first relates to the visible and physical exteriority of a piece of land or a given territory. However, the ultimate cut from the "unknown outside" occurred when the more intangible medium of the air was internalized, too. This makes us understand that not only has the visible naturalness of the world radically changed but also the basic element of human existence has been transformed into a medium to be regulated as needed. Not only has the land and "man's relation to the soil" been changed dramatically but also has the dimension of life itself: air-spaces turn to what Deleuze referred to as "vast spaces of enclosure" (Deleuze 1992, 3).

*

At this point, the sedimentary layers of Crawford Lake offer an overlooked footnote. The radioactive remnants of the 1950s relate not only to the "heavy plow" or the early agricultural practices of domestication but also to an innovation that foreshadowed the anthropocenic world interior already in the second half of the 17th century. In 1660, the scientist Robert Boyle (1627–1692) reported on the success of his recently constructed air pump: A trapped bird – Boyle initially took an injured larch – was enclosed in a glass dome while the air was continuously extracted from it using a pump, to the point where the animal "began manifestly to droop and appear sick, and very soon after was taken with as violent and irregular Convulsions (...) and dyed with her Breast upward, her Head downwards, and her Neck awry" (Boyle 1660, 328). Boyle's experiment seems significant in three ways. First, it concretizes the "spaces of enclosure": spaces whose living conditions can be completely regulated by a simple mechanism and transformed into a hostile or even lethal condition, just as the atomic bomb can internalize the entire planet with a single drop and deprive it of all life. Second, it emphasizes the totality of internalization. Internalization mechanisms are not confined to certain entities but incorporate everything there is. In the wake of the Thirty Years' War, Boyle's "air pump" proved the potential to modify air as the most basic condition of existence and manifested the totality of human dominance and suppression. According to Marijn Nieuwenhuis (2015), with "Robert Boyle's air pump, (…) air transformed into some*thing* through which humanity could now exercise control over life and death" (171). Boyle's air pump allows humanity to entirely internalize nature, granting humans to decide about life and death in a god-like manner. Finally, this total and invisible internalization reveals a third characteristic: if the internalization mechanism refers not only to inanimate matter outside the human body (such as the furrow in the field) but also to the fundamental element of life, this ultimately entails the transformation of the body-psyche itself such that the human becomes human-made. The Anthropocene, as a call for a new geochronological categorization of the earth, implies that the human being not only has started to transform his surroundings but also her- or himself. As long as this biopolitical level cannot find its way into the scientific discourse, any definition of the Anthropocene is insufficient and a provisional rejection as in March 2024 is justified.

It is no coincidence that the painter Joseph Wright of Derby (1734–1797) depicted Boyle's experiment in the darkest colors (Figure 1.1): the people involved in Wright of Derby's 1768 painting appear serious and terrified by the power manifesting before them; in their desire to internalize the outside, they may suspect for the first time that something worrying is going on in the glass dome in front of them. Perhaps they are aware that, on a small scale, a new consciousness and corporeality begin to unfold. Wright of Derby's painting represents the entry into an internalized existence within enclosed spheres of air ("atmospheres") and visualizes what can be defined as "atmospherical consciousness" (Sloterdijk 2009b, 77–78). Such atmospheric consciousness

Figure 1.1 Joseph Wright of Derby, *An Experiment on a Bird in an Air Pump,* 1768, Oil on Canvas. 182.9 × 243.9 cm (72 × 96 in). National Gallery, London.

represents the first politicization of the human body, whose living conditions may be transformed at will. The Anthropocene demonstrates that violent human practices reshape not only the environment's sediments but also the human body's cells, which, with every breath, reveal their internalization and consciousness of floating in a modified and modifiable planetary air space. With these discoveries, Boyle's "air pump" might be considered one of the first biopolitical devices for organizing and altering the human body and the more-than-human environment. Unsurprisingly, Boyle's innovation corresponds with Foucault's procedure of "discipline," whose reign began in the 17th and 18th centuries as well. While humans continue to occupy an individual position, they are fundamentally determined by those in power with atmospheric enclosure and internalization:

> out of a formless clay, an inapt body, the machine required can be constructed; posture is gradually corrected; a calculated constraint runs slowly through each part of the body, mastering it, making it pliable, ready at all times, turning silently into the automatism of habit.
>
> (Foucault 1995, 135)

*

The earliest traits toward disciplinary societies remain entrenched in the more archaic societies of sovereignty. In societies of sovereignty, the sovereign does not yet engage in disciplinary sanction but rather in procedures of punishment:

> The right to punish, therefore, is an aspect of the sovereign's right to make war on his enemies: to punish belongs to "that absolute power of life and death which Roman law calls *merum imperium*, a right by virtue of which the prince sees that his law is respected by ordering the punishment of crime."
>
> (Foucault 1995, 48)

From the God of the Old Testament to Napoleon Bonaparte, societies are marked by fear of their rulers' punishments and rarely decide to openly oppose the sovereign's power, as such opposition could quickly be met with a judgment on one's own life. Exemplarily, Psalm 6 opens with the plea: "O Lord, do not rebuke me in thine anger, nor chasten me in thy hot displeasure." Societies of sovereignty do not yet manifest a biopolitics of internalization but rather a *necropolitics* of unrestricted, inaccessible sovereign power that arbitrarily decides on its subjects' life and death; they are fatalistic societies of separation, enabling people to live free lives as individuals – as long as they know the possibility of punishment. By the end of the 17th century, however, the social structure had changed. Necropolitics transformed into *biopolitics*. The question was no longer on how the sovereign can threaten the subject with death to make them obey, but rather on how the sovereign can *influence* the subject so that they become available to him and do not turn against the sovereign power in the first place. The total separation of exclusion (sovereign vs. subject) is removed and replaced with internalization, appropriation, and manipulation of the subject. The individual ceases to move freely and quasi-limitlessly (under threat of punishment), but becomes, as Boyle clearly shows, an enclosed, internalized subject that moves between different "spaces of enclosure":

> The individual never ceases passing from one closed environment to another, each having its own laws: first, the family; then the school ("you are no longer in your family"); then the barracks ("you are no longer at school"); then the factory; from time to time the hospital; possibly the prison, the preeminent instance of the enclosed environment. (Deleuze 1992, 3)

On this biopolitical level, the tendency toward internalization also reveals some of its fundamental characteristics: "to concentrate; to distribute in space; to order in time; to compose a productive force within the dimension of space-time whose effect will be greater than the sum of its component forces" (3) – all these determinants can only occur in a space of internalization and are no longer concerned with "to rule on death" (3): humans are no longer punished, but disciplined. According to its origins, discipline indicates a different relationship than *Herr und Knecht*: it is a relationship between a student (Latin: *discipulus*) dependent on a teacher. The teacher is "the one who knows" and arranges the student's life; he does not operate despotically but severely. In taking an observing and condemning position, he has the authority to cut the student's word or the student's breath at any moment.

*

These biopolitical aspects of internalization and their techniques for colonizing the outside are now associated with a substantial paradox. One may expect the disciplinary regime of internalization, as shown in Boyle's air pump, to be associated with solid aversion and reactivity. In fact, however, the opposite is true, such that enclosure from the unknown outside at the cost of being disciplined is considered a *desired regime*. Jean-François Lyotard highlighted such a *desire* for internalization by referring to the plight of workers in Britain during the Industrial Revolution when Boyle's apparatus had already been converted into reality:

> If they become the slave of the machine, the machine of the machine, fucker fucked by it, eight hours, twelve hours, a day, year after year, it is because they are forced into it, constrained, because they cling to life? Death is not an alternative to it, it is a part of it, it attests to the fact

> that there is *jouissance* in it, the English unemployed did not become workers to survive, they – hang on tight and spit on me – *enjoyed* the hysterical, masochistic, whatever exhaustion it was of hanging on in the mines, in the foundries, in the factories, in hell, they enjoyed it, enjoyed the mad destruction of their organic body which was indeed imposed upon them, they enjoyed the decomposition of their personal identity, the identity that the peasant tradition had constructed for them, enjoyed the dissolution of their families and villages, and enjoyed the new monstrous anonymity of the suburbs and the pubs in the morning and evening.
>
> (Lyotard 1993, 111)

Lyotard's observation reveals, on the one hand, the "decomposition" of the "peasant tradition" through imposed procedures of manipulation and destruction of one's corporeality, i.e., the shift from necropolitics to biopolitics, or from field to atmosphere. On the other hand, this internalizing decomposition is coupled with pleasure and enjoyment, even if such desire entails one's own and, possibly, the planetary death. Lyotard's observations are not ironic, even if they are not without a certain cynicism. Albert Caraco (1982) presents a similar – even identical – view in his posthumously published *Bréviaire du Chaos*:

> Le mouvement, qui ne s'arrête plus, nous désassemble et nous y consentons *avec délices*, nous approuvons tout bas ce que nous affectons de déplorer, nous *jouissons* de ce chaos insinué dans l'ordre le plus despotique et nous prenons nos libertés de mort au préjudice de nos fins. L'humanité veut pleinement ce qu'elle doit subir.
>
> (27)[2]

*

If there are good reasons for not recognizing the Anthropocene as a geochronological epoch for the time being, one important reason has not yet been mentioned. The violent transformation of the human body and planetary environment cannot be classified in a neutralizing way as "human" but rather as "male." It is, therefore, not appropriate to reflect on an all-encompassing anthropocentrism but on a specific *white androcentrism*, which, consequently, calls for the Anthropocene to be understood as the *Androcene* (ἀνδρός is the genitive of ἀνήρ, which denotes the male adult). It can thus be argued "that the Anthropocene is more accurately the Androcene" and that such a notion is more appropriate "because androcentric technics of phallogocapital are enacted globally in modernity. The global aspect of phallogocapital is postcolonial economic and cultural imperialism rather than universal human responsibility for ecodestruction" (Glazebrook 2016, 62). The tendency toward planetary internalization cannot be assigned equally to all people, men and women alike, but is primarily *masculine*. However, a generalizing emphasis on all men appears to be insufficient, too, as the imperialist component of the "mystery of insulation" is carried out from a specific geopolitical position in which *white men* predominate over people of color of all genders, as well as the Global North over the Global South, and power over marginality. Indeed, the tendency toward internalization does not just imply an escape from a quasi-indeterminate outside, from "a" nature unexpectedly afflicting people – even if this narrative continues to attract significant emphasis in the anthropocenic present. To fully understand the extent of the Anthropocene as the *White Androcene*, we must first examine the predominant concept of nature, namely the meaning of the outside from which humans try to escape. Francis Bacon's 1620 treatise *Novum Organon* offers an essential insight in this regard:

> There is therefore much ground for hoping that there are still laid up in the womb of nature many secrets of excellent use, having no affinity or parallelism with anything that is now known, but lying entirely out of the beat of the imagination, which have not yet been found (...) only by the method which we are now treating they can be speedily and suddenly and simultaneously presented and anticipated.
>
> (Bacon 1855, 105–106, translated in Prior 1954, 351–352)

Bacon's work highlights the imperialist and colonial power toward the more-than-human environment in regard to gender; nature is viewed as innately *female* and so subjugated. He portrays a subservient feminine environment – "the womb of nature" – plundered and deflowered by a violent male cohort – the undefined "we" in the paragraph above. According to Carolyn Merchant (1983), Bacon's claim represents the shift "from nature's servant to its exploiter, and nature from a teacher to a slave" (169), which White (1967) had previously anticipated, too. Therefore, why was it essential to characterize nature as an arbitrarily exploitable female outside that must be suppressed in its agency? Why does nature transition from teacher to slave?

*

We may argue that disciplinary regimes try to wield their disciplinary authority as extensively as possible and do not tolerate any "disruptive third party." For Bacon, this is the existence of an *actual feminine agency* in the planetary environment *that is always outside* and cannot be completely internalized. The "womb of nature" includes an oscillating force, as described in the biblical narrative of creation: "And the earth was without form, and void; and darkness was on the face of the deep. And the *Spirit of God* moved on the face of the waters" (Gen 1:2). The "womb of nature" – the ground "without form, and void" – not only holds "many secrets of excellent use" but also an agency described as the "Spirit of God" or "rûah" [רוּחַ]. It is the third dimension of the biblical God, alongside God the Father and God the Son, yet it is "faceless" and "the most mysterious of the three divine persons" (Johnson 2018, 52). The Spirit of God is a substance rather than an entity, as demonstrated by its wind-like materiality: "Originally, rûah meant 'wind' (...) God's breath" (Cassin 2014, 1014). This aerial substance is etymologically

> allied with female reality, as can be seen not only by the grammatical feminine gender of the term rûah (...) but also by the use of the female imagery of the mother bird hovering or brooding to bring forth life, imagery associated with the Spirit of God in creation (Gen 1:2) and at Jesus' conception and baptism (Lk 1:35 and 3:22). Semitic and Syrian early Christians did interpret the divine Spirit in female terms, attributing to the Spirit the motherly character
>
> (Johnson 2018, 51)

so that it "[is] presuming a genuine side-by-sideness in its participants (...) characterized by mutual trust and responsibility, absorption in common interests, and inclusivity toward others, which may extend even to the stranger and alien" (153). If the unknown outside is excluded using phallocentric instruments such as the "heavy plow" and an insulating regime of discipline is to be built, this must entail eradicating this very feminine spirit that moved on the face of the waters. Rûah is radically opposed to the principles of androcenic transformation, so that the escape from the outside always entails an escape from and a suppression of the active agency of a feminine breath: birth, inspiration, inclusion, trust, responsibility, compassion and care, and carrying: "The Spirit connects human beings, interweaving them in an unexpected manner in diverse structural patterns of life (...) The Spirit comes bringing life 'from all sides'" (Welker 1994, 28). It might be argued that the androcenic tendency to internalize is always accompanied by an exclusionary dominance of rûah's integrative, pervasive, birthing, and inspiring force, as well as an aversion to allowing one's corporeality to be touched by rûah's presence. Indeed, the White Androcene escalates this strong internalizing tendency and control over the planetary interior space, devoid of any feminine aerial agency. It is hardly unexpected that disciplinary atmospheres culminate in the atrocities of the Shoah, which represents the total lack of rûah. The gas chamber is the most extreme internalization and perverted manipulation of human corporeality, demonstrating a still unimaginable power of biopolitical discipline. According to Nieuwenhuis (2018), the "Shoah should not be seen as a historically unique event, but as a continuation of law's ongoing interest in the 'politics of life itself'" (81), that is, as a culmination but also, I argue, as a provisional endpoint of the politics of internalization. Having initiated and experienced World War II, the atrocities of the Shoah, and the testing and deployment of nuclear weapons – politics and institutions of internalization face a *crisis*. The mystery of insulation no longer has any moral justification: "everyone knows that these institutions are finished, whatever the length of their expiration periods" (Deleuze 1992, 4). For those in power, this crisis brings the threat of the rise of feminine meteorologies, which must be prevented at all means at hand. Discipline is confronted with the need to change its rigidity. The power of oppression, internalization, and punishment, evident in Crawford Lake's sedimentary layers, changes and re-emerges in a new form, as the power of *surveillance and control.*

Notes

1 While this proposal was ultimately denied in March 2024 by the *International Committee for Stratigraphy*, the term *Anthropocene* persists – perhaps even more so – as an "uninvited guest" in current debates. In formulating the following thoughts, the Anthropocene is considered more like an event than an officially recognized epoch – retaining its relevance to address the contested space, which white androcenic phenomena continue to occupy. At the same time, this usage acknowledges its indebtedness to earlier critical interventions, for instance those of Donna Haraway and Christina Sharpe, both of which have drawn on the Anthropocene to illuminate the entanglements between race, gender, space, and power.

2 Translated as: "The movement, which never stops, disassembles us and we consent to it *with delight*, we silently approve of what we pretend to deplore, we *enjoy* this chaos insinuated in the most despotic order and we take our liberties to the detriment of our ends. Humanity fully wants what it has to endure."

2 Synchronized Bliss

Around 1950, biopolitical regimes underwent a pronounced shift, aptly characterized by poet Paul Celan as a "breathturn" (Cf. Celan 1995). Following World War II, internalizing practices, progressing from the emergence of the "heavy plow" to Boyle's air pump and culminating in the atrocities of the Shoah, reached a critical point where aerial discipline and the manipulation of both human and more-than-human life seemed no longer possible. The testimony of the gas chambers brought all forms of discipline to an abrupt halt, laying bare an unfathomable abyss. Presumably, the false assessment that the Anthropocene is the onset of a new epoch (while it is, on the contrary, a *termination*) contributed to its scientific rejection. Gas chambers and radioactive remnants do not denote a novel episode but the *inevitable endpoint* of a millennia-long trajectory, reaching its height in the 1950s due to technological development (a notion of development, however, with utmost cynical characteristics). At this juncture, the further enforcement of disciplinary practices becomes untenable: establishing an internalized interior space, where the subversive force of the feminine "rûah" is radically eradicated, and human existence is regulated in an air pump-like manner, is morally and ethically no longer possible. Deleuze noted that "the disciplines underwent a crisis to the benefit of new forces that were gradually instituted and which accelerated after World War II: a disciplinary society what we already no longer were, what we had ceased to be" (Deleuze 1992, 3). Thus, the post-World War II years must be viewed as a period of culmination rather than a beginning, a period, however, where the effects of androcenic practices have become the most pronounced. While the 1950s did not inaugurate a new epoch, they rather marked an endpoint and a subsequent *transitional phase*, a pivotal turning point in human history – and to clarify this transition, the concept of the Anthropocene, understood as White Androcene, proves valid. Celan's notion of a "breathturn" indicates the culmination of preceding internalization practices and the associated *transformation* of biopolitical practices in general. If disciplinary societies were characterized by a perpetual threat of modifying or even extinguishing the human body – akin to the bird in Boyle's air pump – the breathturn of the 1950s signifies a shift toward a post-disciplinary mode of organizing the human body-psyche, a *wholly*

DOI: 10.4324/9781003603672-4

distinct mode of being: a new reality manifesting itself through subtler and indeed gentler forms of discipline.

*

The 1950s are pivotal for biopolitical regimes, as they unfold two opposing tendencies. On the one hand, there exists a sense of powerlessness: established and "universally applicable" power mechanisms can no longer operate in their accustomed manner. Counter-hegemonic forces, such as "rûah" and feminine meteorologies, hover on the brink of disciplinary terrains. Institutional enclosure structures are undergoing replacement, and tendencies toward internalization appear to be waning. Such a condition poses an intimidating challenge to androcenic regimes, akin to intolerable traumatization. However, this sense of powerlessness elucidates merely one aspect of the breathturn; the other aspect assumes a more optimistic hue for hegemonic forces. Even as disciplinary institutions and mechanisms of internalization seem to dissolve, the potential for oppression persists – on a yet unimagined scale. The apparent powerlessness is accompanied by a hidden ingenuity: new practices must be devised, developed, or "designed" to diverge from morally untenable disciplinary practices and continue them with exponential force. Deleuze characterized this new practice as *control*: "'Control' is the term Burroughs suggests for the new monster, a concept Foucault acknowledges as our imminent future" (Deleuze 1992, 4). Those in positions of power no longer enforce discipline but rather control and consent to be controlled. The breathturn of the 1950s necessitates the adaptation of atmospheric control mechanisms: no longer confined solely to technical or physical processes as with disciplinary procedures, the paradigm of control encompasses psychological, social, and economic dimensions. Indeed, the objective is to obliterate the dichotomy between teacher and student and render disciplinary power imperceptible, thus obfuscating its surveilling gaze. The elimination, flattening, and proliferation of the hierarchy of discipline emerge as integral components of these emerging control regimes. No longer is there a solitary, omnipotent observer; instead, each individual assumes the observer role. Gone is the solitary hand commanding the air pump and its subordinate bird; instead, respiration emerges as "breathing together." This novel mode of respiration, coinciding with the demise of disciplinary regimes, signifies a transition from the singular "you" to the collective "we." This shift in pronouns carries profound implications: it signals the termination of a paradigm where individuals or objectified groups were addressed and subjected to discipline. Established disciplinary practices operated within a dichotomy of dominance and subjugation, epitomized by imperatives such as "Youth serve the leader!" or "Do you want total war?" or, in the post-World War II American military lexicon: "These outrages! Your fault!" In each instance, a directive emanates from an unidentified authoritarian position to a designated group, with severe consequences for non-compliance or dissent. This overt objectification of individuals as mere recipients of directives becomes untenable in the ethos of "breathing together." In the 1950s, biopolitical regimes realized that their authority could no longer be maintained through the binary lens of "we vs. you" but call for a decentralization and diversification of power dynamics. The emerging control regimes are premised on the notion that everyone can wield power. This entails dismantling the hierarchical logic of teacher vs. student and transforming the individual into a "dividual." The once inseparable in-dividual becomes *dividual*, assuming a quasi-equal status within an overarching power structure – a floating particle in the global airspace of control. "We no longer find ourselves dealing with the mass/individual pair. Individuals have become 'dividuals' (...) The disciplinary man was a discontinuous producer of energy, but the man of control is undulatory, in orbit, in a continuous network" (Deleuze 1992, 5–6). Humans have transitioned from the disciplinary regime of the "school" to the control regime of the "orbit," where they float as quasi-free, equal particles, liberated from repressive disciplinary constraints. The mantra is "We are all one." The once-regulated world interior space is disintegrating, evolving into a perpetually shifting condition. In 2000, the leftist journal "Terz" proclaimed that "in the control society (...) the institutions have dissolved, but have not disappeared; they have become universal; the prison is now everywhere and nowhere, and everyone is a suspect" ("Kontrollgesellschaft" 2000, translation mine). The ability to float and respire freely in orbit implies constant mutual surveillance and competition; each individual may assume the roles of ruler and oppressor at will. In this aerial regime of control, only the collective "we" exists – an authority no less severe and ruthless than its disciplinary predecessors and perhaps even more so. Indeed, the control regimes espouse a quasi-divine agenda: "I am a God who sees" (Gen 16:13) is applied to everyone. Individual autonomy entails continuous surveillance and control of and by fellow human-particles. Institutional prison walls are rendered obsolete as the established forms of internalization are transformed into universal forms of collective confinement: those who fail to respire "together," "with us," and "in the same pulse" forfeit their right to existence. The atmosphere, once totalized by disciplinary regimes, is now subject to large-scale *standardization*. Controllers and controlled, air providers and air recipients no longer adhere to a hierarchical structure; they now coexist with each other in a more uniform form, their previous exclusions and internalizations blurred. The boundary that once separated the student from the teacher has been erased: the paradigm of control signifies a boundless force of *synchronicity*.

*

The transition from disciplinary to control regimes comes along not with only structural and grammatical shifts (from "we vs. you" to the quasi-inclusive "we") but also with semantic alternations. When it becomes apparent that discipline is no longer tenable, the language of discipline also needs to change. The *language of control* is now characterized by an absence of hostility and a discourse prioritizing collectivism and cohesion. Synchronicity necessitates an intimate and interconnected language divergent from the language of discipline. Dostoevsky suggests that this language embodies bliss and well-being, exemplified in the spatial archetype of control regimes, Joseph Paxton's Crystal Palace of 1851:

> As far as my personal opinion is concerned, to care only for well-being seems to me positively ill-bred (…) *Suffering would be out of place* in vaudevilles, for instance; I know that. In the crystal palace it is even unthinkable; suffering means doubt, means negation, and what would be the good of a crystal palace if there could be any doubt about it? And yet I think man will never renounce real suffering, that is, destruction and chaos. Why, after all, suffering is the sole origin of consciousness.
>
> (Dostoevsky 1960, 31)

As Dostoevsky indicates, the language of synchronization, of collective floating in orbit, is devoid of suffering and uncertainty. Floating together is only conceivable if all entities are organized as smoothly, seamlessly, and transparently as possible. In this context, philosopher E.M. Cioran identifies a burgeoning "indolence," "subtlety," and "repletion" (Cioran 2010, 116–117) within the language of bliss. However, it is suffering and doubt that truly invigorate: "Under the goad of pain, the flesh awakens" (Cioran 2010, 42). If Dostoevsky perceives Paxton's Crystal Palace as the embodiment of well-being, this spatiality, at the same time, signifies the absence of suffering and, thus, of life itself. Instead, it epitomizes a sentimental, blissful morality devoid of pain and doubt.[1] In short, we might say that control regimes deny bodily liveliness or, more accurately, humanity itself. If collective respiration necessitates the rejection of individual corporeality and its "debility" (Cioran 2010, 100), then such spatiality must also eschew acknowledgment of the debility of others. A communal existence is only possible at the expense of a global "indolence" that eradicates the excesses of individual suffering and compassionate alterity. However, the insidious aspect of the language of control lies particularly in its ostensible semantics of care and concern. Communal language and synchronized breathing patterns are particularly evident in overarching, heart-felt narratives, as seen – perhaps inadvertently – in the poetry of pacifist author Barbara Deming, whose work is currently being revisited by activists:

> Our own pulse beats in every stranger's throat, And also there within the flowered ground beneath our feet, And – teach us to listen! – We can hear it in water, in wood, and even in stone. We are earth of this earth, and we are bone of its bone.
>
> (Deming 1984, 247)

Within such language, the human body-psyche becomes inscribed into and absorbed by more-than-human phenomena: we are all interconnected, humans as well as water, stone, wood, and earth. This underscores the all-encompassing ethos of the new discipline: the call for blissful, sentimental, and enthusiastic coexistence. When philosopher Eva von Redecker advocates a "venture of remaining soft," Cioran's observation of burgeoning "indolence" is particularly close;[2] indeed, many contemporary posthumanist perspectives can be considered indolent. While these viewpoints advocate communal unity "with all that is," they also exemplify chilling manifestations of coldness and brutality. Beneath the grand declaration of coexistence lurks compulsion and sheer violence: if you refuse to coexist with us, you will perish.

*

The brutal language of bliss, with its seamless and quasi-inclusive transition between subject and object, is also marked by its deceptive intent. Control regimes designed to thwart disciplinary measures are inherently geared toward presenting a different facade. Their fundamental propagandistic technique is as simple as insidious: an abundance of linguistic softness suggesting the absence of severity and discipline. However, a closer examination reveals quite the opposite; the veneer of gentleness merely serves to conceal underlying violence. Cioran posed a fundamental question in this context: "What use displaying your failure when you can feign prosperity?" (Cioran 2010, 109). The key distinction between disciplinary and control regimes lies in the undisclosed and positive promotion of their power mechanisms. In everyday life, these mechanisms are manifested in the proliferation of feigned "cooperative" and "considerate" forms of thought and action.[3] The language of bliss has little to do with genuine human fortune but is bound to doom and demise. As everyone floats together in the illusion of their supposed freedom, they become entirely observable and controllable. Moreover, this omnipresent visibility facilitates the punishment and sanctioning of those who deviate from blissful coexistence in both real and virtual realms. Everything is visible to everyone, and today's victim could become tomorrow's persecutor. Thus, assuming freedom when floating across species boundaries proves to be a mere illusion. In a world where everything is conceivable, where my pulse resonates in the earth's flowers, and where everyone is joyously connected to everything, true freedom is nonexistent. A single, synchronized breathing rhythm becomes imperative, allowing no room for deviations. Floating together necessitates the becoming-synchronized with the universal pulse.

*

As the human organism transitions from an individual to a "dividual," it becomes more of a number than a person. It may be manipulated at will but must maintain its numerical identity to continue existing in the global crystal palace. The synchronizing agenda of control can also be expressed biologically, as Lynn Margulis and Dorion Sagan do:

> At every level, from microbe to planet, organic beings use air and water or other organic beings to build their reproducing selves. Local ecology becomes global ecology. As a corollary, and in spite of English grammar, life does not exist on Earth's surface so much as it is Earth's surface. Life extends over the planet as a contiguous, but mobile, cover (…) Organisms are less self-enclosed, autonomous individuals than *communities of bodies* exchanging matter, energy, and information with others. Each breath connects us to the rest of the biosphere, which also "breathes," albeit at a slower pace.
>
> (Margulis and Sagan 2000, 23)

Margulis and Sagan accurately articulate the abstraction of universal floating together, asserting that "life extends over the planet as a contiguous but mobile, *cover*." Consequently, otherness and difference have vanished not only on an individual but also on a terrestrial scale: life appears on a flat cover of exchange. Humanity has regressed into superficiality. The only necessity for the human being is to maintain its cellular breathing with all others and at no cost to make a mistake, express a doubt, pain, weakness, or resistance:

> The control society is not a hierarchically ordered, homogeneous-totalitarian whole with a rigid ideology; it does not see itself as a moral institution; the collapse of values has also affected it. In principle, it does not care if humans consume illegal drugs, but it does care where they do it. "You should not inject heroin in front of shopping centers and train stations!" is the order of the day. No punishment can be derived from this relativism of values.
>
> ("Kontrollgesellschaft" 2000, translation mine)

*

Deleuze accurately highlights that floating within the dividual cover, despite (or perhaps because of) its blissful condition, should be considered as the "brashest rivalry (…) that opposes individuals against one another and runs through each" (Deleuze 1992, 5). In his observation, he also references Paul Virilio, who identified the phallocentric qualities of this world-embracing cover as "ultrarapid forms of free-floating control" (Deleuze 1992, 4). Assuming that "speed has always been the advantage and privilege of the hunter and the warrior" (Virilio 1994, 19), a direct link between speed and masculinity emerges, which, incidentally, laid the groundwork for the brutal agency of control regimes.[4] Life within the flat cover also unfolds a gender- and race-specific deception. Although rights of women, people of color, or people with disabilities seem to play significant roles in contemporary discourse, these objectives are not immune to inherent indolence either. Believing that the language of bliss acknowledges the feminist, racial, or inclusive agenda is delusive; on the contrary, their obsession with speed and ongoing rivalry renders any reference to the other in her/his/its specificity impossible. The present indolence refers to the rights of the other only on a superficial, cliché-like level, just as it would be simply wrong to suggest that, e.g., softness and femininity are synonymous. The rise of the "soft girl" phenomenon on social media since the late 2010s indicates that such incompatibility has lost its rationale, and a nostalgic and reactionary form of femininity is currently celebrated.[5] Advocating feminism with the paradigm of softness must be unmasked as fundamental deceit, however ambitiously perpetuated by control regimes. In reality, the femininity espoused by the language of bliss only superficially aligns with a feminist agenda; since the paradigm of softness lacks any unsettling layer of meaning, Audre Lorde's well-known feminist call is deliberately misappropriated and proclaimed on a mere literal level: "Caring for myself is not self-indulgence, it is self-preservation and that is an act of political warfare" (Lorde 1988, 131). The soft superficiality of control regimes raises doubts on whether "caring for myself" truly aligns with a political or counter-hegemonic agenda. Nigerian author Désọ́lá Ọlálẹ́yẹ recently remarked on this issue:

> I recognize the value of encouraging black women to care for themselves and cultivate a lifestyle that enables inner peace. But I question if a soft lifestyle, in its common expression, bears the same liberatory politics as Lorde's feminist call to nurture the self. Lorde does not remove her awareness of the need for social transformation from her promotion of self-centeredness.
>
> (Ọlálẹ́yẹ 2023)

Ọlálẹ́yẹ's observation underscores the *lack of subversive force* in the quasi-feminist agenda pursued by control regimes. Despite the pervasive tendency of becoming soft, such an agenda remains devoid of any genuine feminist ethos, with gender differences being suppressed in favor of an aggressive yet hidden masculinity centered on speed and progress. Deviances and differences vanish within the superficial cover of collective breathing, with everything and everyone operating as "transparent" numbers or exchangeable air particles. The breathing patterns of control regimes are entirely synchronized, concealed behind veils of softness, apparent kindness, and bliss. Feminist agenda has never been further from its intended target than in controlling propaganda, making rebellion against such a form of oppression nearly impossible – as those in power would reply with a quasi-surprised expression: "It's all there – more than ever before! Why are you rebelling? The borders have fallen! Be free!" The tragedy lies in that the more individuals are told they can freely float within the universal orbit, the more they cling to this false narrative, only realizing the truth of their actions afterward. "Were not our hearts burning as he talked with us on the way? (Lk 24:32) – in a contemporary translation: "Were not our hearts burning as the language of bliss began to choke off our breath more and more?"

*

Margulis' and Sagan's proclaimed reality of the flat cover manifests as a space devoid of substance. A third illusion emerges: the orbit, identified by Deleuze as the working sphere for control regimes, is not a multifaceted spatiality but a smooth and monotonous construction – a flat plane lacking dimensionality and alternative coordinates and directions. Byung-Chul Han underscores this notion, identifying the superficiality of the blissful sphere of control both on a large scale and in everyday details, such as smartphone displays:

> The smooth characterizes our present. Do you know the G Flex, a smartphone from LG? This smartphone has a special coating: If scratches occur, they disappear after a very short time, so it has a self-healing skin, almost an organic skin. This means that the smartphone remains completely smooth. I ask myself: why do a few scratches on an object bother you? Why is this striving for a smooth surface? (...) This smooth surface of the smartphone is a skin that is not vulnerable, that resists any damage (...) You don't want to be vulnerable, you shy away from any injury or being injured.
>
> (Han 2014, translation mine)

Interestingly, the reality of control regimes indeed mirrors the technology of smartphone displays. This flat technology is interesting for the characterization of the breathing rhythms of control for several reasons. Besides its smoothness and quasi-invulnerability, it reveals a more considerable existential aspect: the flat display is an externally controlled and highly addictive apparatus. Scientists have established a correlation between excessive display usage and physiological changes such as blood volume pulse, skin conductance, peripheral temperature, and respiratory response. Recent studies indicate alarming rates of internet addiction among Americans, with 41% showing probable addiction and 15% exhibiting strong addiction (Cf. Khubchandani et al. 2021). When extrapolated globally, these figures suggest a significant amount of the world population experiences altered respiratory rhythms due to excessive display exposure: "the (...) RESPR [respiratory response] of high-risk IA [internet addiction] abusers were respectively (...) stronger than those of low-risk IA abusers" (Lu et al. 2010, 371). These findings exemplify the order of "breathing together" in its most immediate and perhaps unsettling manifestation. Breathing ceases to be an individual, emotionally charged, and com-passionate relation between the human body-psyche and its surroundings. Instead, it becomes a reflexive response to the stimuli emitted by flat, invulnerable displays. This phenomenon illustrates a profound connection between Margulis and Sagan's biologistic perspective of life as a "cover," Cioran's notion of "indolence" and absence of suffering, the promoted language of bliss, and physiological evidence of altered breathing patterns due to display-mediated stimuli – the smartphone display as the technological equivalent of the biologistic "cover." Control regimes thrive on mechanisms of synchronization, forcing individuals to conform to a flat grid life, modify their breathing patterns, and propagate the illusion of floating freely in space. This realization is both frightening and deeply pessimistic. Deleuze offers a somewhat optimistic perspective in his "Postscript on Societies of Control":

> For example, in the crisis of the hospital as environment of enclosure, neighborhood clinics, hospices, and day care could at first express new freedom, but they could participate as well in mechanisms of control that are equal to the harshest of confinements. There is no need to fear or hope, but only to look for new weapons.
>
> (Deleuze 1992, 4)

Recently, the task of discovering or forging "new weapons" has become increasingly daunting. The pervasive atmosphere of "breathing together" and relentless propagation of bliss severely cuts, if not truly obstructs, the potential for resistance and the creative cultivation of counter-hegemonic practices and tools. Entrapped in dependency cycles, individuals have grown accustomed to the narrative of collective freedom and synchronizing softness that obliterates all differences. The language of bliss no longer provides the linguistic scaffolding or vocabulary necessary to articulate "new weapons." Simultaneously, it is imperative to consider that control regimes have always been armed despite the veneer of peaceful coexistence they present. The totality that characterized disciplinary regimes and their associated dichotomies ("we" vs. "you") is conspicuously absent in the realm of control regimes. The mantra "we are all one" conveys an alarming degree of societal brutality. Humans have relinquished their discipline and instead adopted a state of indifference, softness, synchronization, and indolence. The insidious nature of these processes lies in the fact that indolence no longer necessitates sovereign sanction ("we") but rather manifests as a pervasive control exerted by all, driven by the fear of a potential loss of personal bliss and dependence thereon. Herein lies the fundamental contrast to disciplinary societies: the prevalence of *dependency* on the flatness of technological displays and terrestrial covers, coupled with the immediate distribution of bliss-inducing stimuli. These stimuli also propagate the illusion of the transfer of smartphone display's purity and invulnerability onto the human body-psyche. Therefore, it is unsurprising that a dependency on a totalitarian system easily takes root when it floods its followers with positive stimuli related to the false assurance that clinging to its synchronized breathing patterns renders them quasi-invulnerable. A 2022 newspaper advertisement seductively proclaims the need for "a 'reset' to bring life energies back into balance," calling individuals to cleanse and strive for resilience collectively – and to take over control over each other.

Notes

1 Peter Sloterdijk elaborated on this point extensively in his book *The World Interior of Capital*. Dostoevsky not only took a critical view of Paxton's London Crystal Palace but also its further discussion in Nikolay Chernyshevsky's 1863 novel *What Is to Be Done?*: "He [Chernyshevsky] had announced, with consequences that extended all the way to Lenin, the birth of the New Human Being, who, having found the technological solution to the social question, would live with its kind in a communal palace of glass and metal – the archetype of living communities in the East and the West. Chernyshevsky's culture palace had been designed as a climatized luxury shell in which there would be an eternal spring of consensus. Here the sun of good intentions would shine day and night, and the peaceful coexistence of all with all could be taken for granted. Boundless sentimentality would characterize the internal climate, and an overstretched humanitarian domestic morality would result in a spontaneous empathy of all with the fates of all" (Sloterdijk 2013, 170).

2 What is particularly interesting about Von Redecker's call is the image of letting humans float through oceans without boundaries and weightlessness – without separating land and sea, human and animal, with the claim of establishing a more-than-human continuum: "The counter-concept to this side of hardening is the very dazzling and elusive image of a mimetic reflection. Mimesis can mean various things, such as imitation and copying. However, the core of this term in Adorno and Horkheimer is different, namely, it refers to something like devotion to the other or absorption in others. The venture of remaining soft is also always the venture of losing oneself. Earlier, we spoke of the logic of ownership, which drastically separates subject and object, and it can be said that the idea of mimesis represents a counter-model to this, in which subject and object merge, are in contact, are also in an unconscious verbal struggle for demarcation. Due to its softness and constant change of form, I considered the octopus a beautiful symbolization of this idea of mimesis" (Von Redecker 2022, translation mine).

3 Hannah Arendt (2006) aptly pointed out the power of language regulations and concealed, emphatically positive semantics using the example of Adolf Eichmann: "Furthermore, all correspondence referring to the matter was subject to rigid 'language rules,' and, except in the reports from the *Einsatzgruppen*, it is rare to find documents in which such bald words as 'extermination,' 'liquidation,' or 'killing' occur. The prescribed code names for killing were 'final solution,' 'evacuation' (*Aussiedlung*), and 'special treatment' (*Sonderbehandlung*); deportation (…) received the names of 'resettlement' (*Umsiedlung*) and 'labor in the East' (*Arbeitseinsatz im Osten*)" (85).

4 Interestingly, such a link between speed and masculinity is also echoed in aggressive artistic movements such as Italian Futurism. Cf. two excerpts from Marinetti's Futurist Manifesto from 1909: "We affirm that the world's magnificence has been enriched by a new beauty: the beauty of speed (…) We will glorify war – the world's only hygiene – militarism, patriotism, the destructive gesture of the freedom-bringers, beautiful ideas worth dying for, *scorn for woman*" (Marinetti in Apollonio 2001, 21–22).

5 Donna Haraway's feminist call for becoming-compost, signifying a becoming-soft in the sense of becoming one with the earth, strikes a similar note: "We are not posthuman; we are compost. We are not homo; we are humus. We are terran; we are earthlings; we are many; we are indeterminate. We bleed into each other in chaotic fluid extravagance" (Haraway in Braidotti and Hlavajova 2018, 79).

3 Anesthesia and the End

The enforced "breathing together" within control regimes is characterized not only by the absence of suffering and pain but also by dependency on the language of bliss and the illusion of invulnerability. As noted by Cioran, such compelled communalism is further marked by a form of indolence that manifests as numbness, painlessness, and absence of compassion with the human and the more-than-human other. However, by the late 2010s, this coexisting insensitivity had become even more pronounced, such that the immunity had transformed into a *lack* of engagement and perception altogether. In this context, Byung-Chul Han identifies an emerging apathy toward reality: "The post-factual age of fake news and deepfakes leads to an apathy towards reality, even an anesthesia that protects us against reality. It is only when a painful shock of reality strikes us that we are able to emerge from this anesthesia" (Han 2021, 32). In this "post-factual" age, which has been solidifying for approximately eight years (roughly coinciding with the commencement of Trump's first presidency on January 20, 2017), there is not only a demise in the perception of pain and vulnerability (in oneself or others), but a disinterest in reality altogether. While the spatiality of the cover is essential in the conception of control regimes, as we have seen in Margulis and Sagan, with the recent development of control regimes, even interest in this planetary cover and its blissful configuration appears to have faded. In *The Symbiotic Planet*, Margulis remarks on the extent of human activity within the planetary cover:

> To me, the human move to take responsibility for the living Earth is *laughable* – the rhetoric of the powerless. The planet takes care of us, not we of it. Our self-inflated moral imperative to guide a wayward Earth or heal our sick planet is evidence of our immense capacity for self-delusion.
>
> (Margulis 2001, 143)

In this statement, Margulis anticipates apathy toward reality, emphasizing that humans within contemporary control regimes have no individual agency and must only fulfill their symbiotic and synchronized responsibilities. However, with this seemingly casual observation of emerging anesthesia,

DOI: 10.4324/9781003603672-5

control regimes have entered a phase of even more profound oppression: On the one hand, humans realize their lack of influence and their inability to deviate from synchronized breathing patterns. On the other hand, Margulis' observation suggests that even if humans were granted influence, they would be impotent: "The human move to take responsibility for the living Earth is laughable." Here, the language of bliss takes on a decidedly bleak tone, expressing not only apathy toward reality but also apathy toward oneself and toward the other. Margulis asserts, "the planet takes care of us, not we of it. Our self-inflated moral imperative to guide a wayward Earth or heal our sick planet is evidence of our immense capacity for self-delusion. Rather, we need to protect us from ourselves" (Margulis 2001, 143). Margulis suggests that assuming that humans can effect change on their own is a downright delusion: they lack both authority and capability; it seems sufficient for them to disengage from themselves, renounce themselves, and submit to the unhindered synchronizing operation.

*

Whereas in established control regimes respiration has been reduced to a flat and synchronized cover level, breathing is *entirely at risk* in this most recent advancement of reality apathy. This *post-anthropocenic* phase could be characterized as a transition from resistance to anesthesia, from illusion to desired and enjoyable self-destruction. The cover's spatiality loses territoriality and dimension and flattens to a mere technical infrastructure. Humans no longer breathe in a two-dimensional network of positive commonality but in one-dimensional linearity attached to single breathing machines. These devices are significantly less noticeable than their name implies. Byung-Chul Han detects such breathing machines not just in the flatness of the smartphone display but in the ephemerality of everyday digitalization, entertainment, and social media: "Digitalization also increasingly reduces resistance, makes antagonistic opposition, the anti, the anti-bodies, disappear. The ubiquity of the *like* leads to a numbness that undermines reality. Digitalization is anaesthetization" (Han 2021, 32). From a psychoanalytic perspective, one might adopt Deleuze and Guattari's framework, wherein in mechanical one-dimensionality "it is the social repression of desire or sexual repression – that is, the stasis of libidinal energy – that (…) engages desire in this requisite impasse, organized by the repressive society" (Deleuze and Guattari 2000, 118). The term "impasse" succinctly captures the one-dimensionality of reality apathy: breathing at a *terminus* or even breathing within a *trap*. Communally controlled breathing transforms into isolated, fully mechanized, or digitalized breathing. Humans relinquish their remaining respiratory agency – willingly and even with a sense of relief. No longer are they within a synchronized cover orchestrating their breaths into blissful rhythms, but rather external to any perceptible structure. The deceit of the Post-Anthropocene lies in its proclaimed departure from any semblance of totality: there exists neither a spatial atmosphere nor a flattened cover, only a contextless control center wherein humans are attached to machines without even perceiving it – which is, cynically, the culmination point of totality. It is a terminus where the pathways of the former cover converge and decelerate. Through their escalating apathy toward themselves, humans permit themselves to be induced into an artificial slumber, relinquishing control of their respiration to machines and algorithms that redefine their breath. Whereas the "dividuals" of traditional control regimes were still precursors to individuals, humans are now entirely renouncing and surrendering their subjectivity by voluntarily abdicating their respiratory function. This is what is referred to as *efficient or optimized breathing*.[1]

*

In *Breath of Proximity*, Lenart Škof claims that the individual respiratory function plays a pivotal role in the formation of subjectivity: "Breathing, not merely air, is what endows humanity with subjectivity and provides the foundation for its psychological and spiritual dimensions" (Škof 2015, 134). It seems, thus, profoundly troubling if humans now become numb to their breathing and willingly relinquish it to machines and algorithms. Departing from the illusion of equal and communal respiration, it marks yet another, highly dramatic, step that humans are taking: recognizing that it is much easier to not breathe for oneself (within a synchronized rhythm) but to be anesthetized and allow oneself to be breathed for. Humans potentially abandon one of their last remaining tools and, consequently, their capacity to humanely and spiritually engage with their environment. While synchronized bliss was once characterized by a phallocentric competition and contempt of potentially differently breathing individuals, global anesthesia reveals itself as the termination of any ethical obligation toward oneself and toward one another. Peter Bojanic underscores this point when he asserts that "opening for the other begins with breathing, with the lungs (lungs are the real subject and the real beginning, a beginning before any other beginning)" (Bojanic in Škof 2015, 135). This observation echoes Emmanuel Levinas, who suggests "that the subject could be a lung at the bottom of its substance – all this signifies a subjectivity that suffers and offers itself before taking a foothold in being. It is a passivity, wholly a supporting" (Levinas 1991, 180). The lung can be conceived as an *ethical organ* that embraces the ambient otherness lying beyond it without anchoring it in deterministic categories of being:

> There is exposure without assumption, which would already be closedness. That the emptiness of space would be filled with invisible air, hidden from perception, save in the caress of the wind or the threat of storms, non-perceived but penetrating me even in the retreats of my inwardness, that this invisibility or this emptiness would be breathable or horrible, that this invisibility is non-indifferent and obsesses me before all thematization, that the simple ambiance is imposed as an

> atmosphere to which the subject gives himself and exposes himself in his lungs, without intentions and aims.
>
> (Levinas 1991, 180)

Here, Levinas articulates the fundamental aspects of a subjectivity oriented toward others, fundamentally defined by its lungs and their function. It is not only the respiratory rhythm that constitutes individuality, but also the fact that the human body carries within itself an organ that absorbs the other in a passive, almost selfless manner and allows itself to be permeated by this other. Levinas' assertion holds significant implications beyond the blissful language of coexistence: through the lungs, the other is carried, possibly in all its horror and threat. Levinas elucidates that the lungs serve a purpose beyond mere romantic coexistence.[2] Through the lungs, we might suggest, "the subject does not find itself any place even in its own volume (…) It opens in space but is not in-the-world" (Levinas 1991, 180). The human subject finds itself withdrawn from its primacy and position, carrying within itself the potential to expose itself to the other.[3] According to Levinas, it is not merely a possibility but even the ethical obligation of every human not to withdraw from the other, not to abandon her/him/it, but to strengthen her/his/its weakness, to take responsibility for her/him/it, and to leave aside any destructive power of self(ish)ness. Significantly, human life is impossible without breathing, and humans indispensably require their lungs to sustain their vitality; simultaneously, we see that being humane fundamentally depends on this lung function – i.e., on exposing oneself to the other. If breathing finds itself at a one-dimensional impasse, attached to a breathing machine, willingly abolished, then breathing fundamentally relinquishes the potential of engaging with the other, and its potential of being humane.

*

The capacity to breathe and to welcome the other into one's corporeality is no longer possible in the current state of reality apathy: access to one's lungs has been rendered inaccessible; the ethical organ has been surrendered to mechanical control; the once flat cover sites of control regimes have been reduced to mere interfaces of anesthesia; the human being has been reduced to a mere junction point where the tube of the breathing machine can be attached. However, this attachment occurs without force; instead, apathy provides solace and relaxation for the human subject. It is a relief that, although hidden, acts as an incessant numbing. The infrastructure of 8 billion breathing machines operates ceaselessly; autonomous breathing has become unimportant, and the lungs are now deemed an inefficient tool. The Post-Anthropocene no longer entails the excessive transformation, discipline, or control, but rather the flattening of everything, rendering it efficient, optimizing it, and ultimately depriving humans of their breath and breathing itself. Humans fade away and willingly approve suffocation: with the algorithmic and mechanized breathing to which they have become attached, a size of violence is inflicted upon them that has escalated to almost unimaginable proportions – yet materialized in its actual dimensions as inconspicuously, as pleasantly relaxing as possible. In reference to Klossowski, Lyotard characterizes such emerging reduction of subjectivity as "tender cruelty" (Lyotard 1993, 74):

> The question of violence, then, is posed with a quite new simplicity, it is displaced into all indifferent and tender cruelty: "Yet what can we say of the violence of one breath towards another? Can the latter condemn the first for having destroyed its fragile habitation, when it should be free of all need to remain the same? Relieved of the need to remain the same, the victims' breaths merge with the victimizers once they see them coming. The latter seem not to know the shame of seeing themselves thus welcomed by the others. No accusations or regrets on either side, and no forgiveness. There is no moral atonement here, and such could hardly be required. A violence of another order is born of our condition: it is effected by means of a *total indifference*. It is this indifference itself: and it leaves no trace, which is the worst form of violence!" The suppression of memorable and mnesiac bodies permits the interpenetration of intentions, that is to say their abolition for the benefit of anonymous intensities.
>
> (Lyotard 1993, 74)

Lyotard's insight encapsulates the attributes of reality apathy: A breathing regime that inflicts violence upon the other through its breath of mechanized ventilation; violence acted upon through fundamental indifference or indolence, which numbs subjectivity and vitality; a fusion of the victims with the victimizers, as the technological breathing infrastructure loses any differentiating perception in its contextlessness and abstraction; and finally, the recognition that this is the worst form of violence precisely due to its tenderness. A wall control panel in my hotel room serves as an illustration of this post-anthropocenic apathy and infrastructure of tender cruelty (Figure 3.1).

The device is a control for a ceiling fan whose rotation can be adjusted with a small wheel. It is crucial to note that the control panel is labeled "Best @ir®" – a control purportedly capable of generating the "best air" through its digital mechanism. This small device vividly illustrates "reality apathy" and the inherent cruelty in an almost naive manner: No longer is it necessary to breathe independently – regardless of how regulated and synchronized it may be – and seek air autonomously, thereby also engaging with the aerial otherness that permeates one's corporeal existence. Instead, the device shows that a mere control panel and a five-part scale are necessary to provide this air and its conditions.

*

Figure 3.1 Control Element for a Ceiling Fan. Own photograph, 2023.

Breathing, as expounded by Levinas, inherently entails turning toward the other. Due to the breathing pulse of the lung ventilation, the lungs are precisely exempt from receding behind the totality of the self: in breathing beyond regulation or apathy, it is conceivable for humans "to open oneself as space, to free oneself by breathing from closure in oneself" (Levinas 1991, 180). However, this act of breathing is also always accompanied by a movement – a *being moved by the other*, who alters and transcends the limitations of one's self. Such a process of being moved presupposes the mobility of humans and their capacity to be spontaneously drawn and permeated by the other; Levinas refers to this as a "restlessness of respiration" (Levinas 1991, 180). The control panel of the "best air" inhibits such restlessness and responsiveness to the other and fundamentally alters one's ability to respire. In the present apathy, any restlessness vanishes, and with it, the presence of the other, and the necessity of humanity. The only movement that remains is using a device measuring approximately 13 by 9 centimeters: the "best air" becomes instantly accessible, and the human is attached, "plugged in," with their suffocation already pre-programmed. The stark contrast to the suicidal fascination in disciplinary societies, as elucidated by Lyotard in Chapter 1, now lies in the fact that the disciplined human still harbored a goal while enjoying her/his demise: enhancing productivity, gaining recognition, fostering competition, acting "blissed," becoming dependent. Conversely, in "reality apathy," all goals have become inconsequential since, in Margulis' words, the belief in achieving any goal is merely a hubris of human existence. A new human emerges that no longer harbors any interest in the surrounding world and has relinquished all concern. Biblically speaking, the Pauline assertion that a new human is being fashioned in the image of their creator could be invoked – the new human as the likeness of a controlling, indifferent deity: "Seeing that ye have put off the old man with his deeds; And have put on the new man, which is renewed in knowledge after the image of him that created him" (Col 3:9b–10). However, such a perspective is valid only superficially. On closer examination, it becomes apparent that in the Post-Anthropocene, the new human will cease to be human altogether; she/he will have forfeited all humanity; she/he has relinquished breathing and attached herself/himself to the breathing machine: indeed, the worst form of violence.

*

Such development signifies the utmost catastrophe for humanity,[4] opening the doors to societal and planetary barbarism. Author Elif Shafak already warned in late 2021: "Indifference is the biggest threat to humanity right now" (Shafak 2021) – a statement shared by numerous observers. Indeed, it must be acknowledged that this threat is not a prospect but already exerting a global influence. Indifference is perceivable not only in the atrophy of the ethical function of the lung – entailing the demise of all humanity and openness to others – but also encompasses the planetary condition as a whole. This elucidates why, alongside "apathy toward reality," apathy toward the world around us and increasing indolence toward profound planetary events are equally perceivable. With these observations, we can only forecast bleak scenarios, suggesting that human and planetary catastrophes are imminent and ready to further materialize in the forthcoming decades. Climate activists have already embraced such narratives of doom – perhaps more for brevity than for dispossessing control regimes and breathing protocols. However, what predictions can be ventured beyond populism in a practice of deconstructing the existing anesthetic apparatus?

*

E.M. Cioran confronts us with the assertion that "breathing itself proceeds from an implicit fanaticism" (Cioran 2010, 73). In the context of breathing within "reality apathy," this assertion proves especially true: the delegation of the breathing function to digital machines is founded on fanaticism, characterized by a violent fixation on efficiency, optimization, resilience, invulnerability, homogeneity, and standardization. It is constituted by indifference and sustained by the "joy of inanity" (Cioran 2010, 110). However, the impending apocalypse of apathy, prophesied to culminate in the near future, also unveils a *second way* that might steer us away from impending catastrophe. Traces of such a way are not discernible in efforts to combat pessimism but rather *in engaging with it and immersing oneself in it*. Cioran's pessimistic aphorisms, for instance, do not merely plunge us into the abyss but rather present a last remaining opportunity to go beyond annihilation. Alenka Zupančič recently advocated a similar line of thought: in her view, it is imperative to construct a reality that no longer cowers in fear of the impending downfall but rather focuses on how that very reality, which would inevitably be lost through the anticipated imminent end, *might first come into existence*: "the true choice is between losing it all and creating what we are about to lose: only this could eventually save us, in a profound sense" (Zupančič in Žižek 2018, 112). Such argumentation is not about indulging in a global doomsday scenario, which would only escalate the fanaticism of indifference and apathy. Instead, Zupančič suggests that the imperative lies in rethinking, with the last remaining *alveoli*, those practices that have long been condemned; in grasping what the breathing apparatus has made all too convenient for humans; in resetting the control mechanism to "zero" and seeking the "best air" beyond digital algorithms. However, this can only be achieved with an attitude of *affirmation* of the existing, in a conscious embrace and acceptance of the dark and uncanny; with Cioran, it becomes evident that such a revived breathing practice cannot emerge from negating prevailing fanaticism but only by *recognizing and taking seriously* its foundational aspects. This *paradigm of affirmation*, of engaging with pain, trauma, and the abyss of the present, breaks away from prevailing anesthesia and apathetic breathing practices; it consciously severs the breathing machines' tubes; it refuses to be perpetually anesthetized by the bliss of reductionism; it transcends the digital realm; and ultimately, it signifies not self-annihilation but the *sole possibility* of continued existence: "The air gone, we should still breathe; but we should immediately smother if the joy of inanity were taken from us" (Cioran 2010, 110). In this statement, Cioran suggests two breathing regimes: a predominant regime of digitalization, mechanization, and immobility and another yet hidden regime beyond the hegemonic powers. Given that apathy continues to sustain human existence, humans would promptly suffocate once this regime is gone. If they, however, discern that there is a breath *beyond or before* apathy, even if such a breath is not yet manifested and can only be perceived intuitively, this establishes the precondition of breathing even in the absence of prevailing breathing regimes; it entails breathing an *unfamiliar air* that has not yet been appropriated or transformed – air composed of the known and the unknown.

*

The axioms of François Laruelle's recently more prominently received non-philosophy can be used to theorize such a different air. Non-philosophy (Cf. Laruelle 2017) is not the negation of philosophy and does not propose a new alternative to the often-proclaimed end of philosophy either. Rather, it describes a way of thinking that can be defined in analogy to non-Euclidean geometry, which does not imply the abolition of geometry but rather its *extension* to a space in which previously unknown and alternative geometries become possible. Although non-philosophy is fundamentally based on established philosophical theorems, a change or mutation of what has previously determined the philosophical canon can be noted here. Non-philosophy introduces a way of thinking that questions and lifts the supremacy of philosophy and can be best described as equal, democratic thinking. Such thinking is not restricted to philosophy alone but extends to everything thinkable, offering the possibility of allowing existing formalisms to mutate and transform: "This is an inversion that concerns a reversal of old hierarchies, but through a formulation of a new type of primacy without relationships of domination; without relations in general" (Laruelle 2012, 232). Jonathan Fardy, in his captivating analysis of Laruelle's thought, summarizes that "non-philosophy is firstly a critique of the authority and power of philosophy and an affirmation of the thoughtfulness, criticality, and creativity of other forms of intellectual expression that do not bear philosophy's authority and approval" (Fardy 2020, 13). In analyzing the hegemonic forms of breathing and the indifference-promoting mechanisms of controlled air, the medium of air can now also undergo a non-philosophical investigation – possibly opening up a second breathing path. From a non-philosophical perspective, air transforms into *non-air* – becoming a life-giving medium no longer manipulated and infused into humans. Non-air signifies the mutation of what hegemonic discourses had classified and defined as air so far; an aerial continuum referred to by Laruelle as "the Real" or "the One" – existing *before* any attribution, comprehensibility, or determination: a continuum within which all classifications and determinisms are immanent, yet remaining indeterminate by nature: "Laruelle axiomatically insists that the Real is radical immanence itself. The Real is radical immanence itself of which thought itself is a part (...) There is no position or place exterior to the Real from which any thought (philosophical or not) can survey and grasp the Real" (Fardy 2020, xiii). Non-air could never be determined or altered externally; rather, it encompasses everything other beyond the human realm (yet equally permeating them). Non-air eludes manipulation, classification, or alteration, perpetually remaining radically foreclosed. It embodies a multiplicity including all life yet existing before determinisms, hegemonies, or ideologies – with this "before" encompassing both temporal and conceptual dimensions: Air transcends traditional representations and goes beyond any form of categorization. It exceeds the confines of a clearly defined gas mixture, encompassing not only a series of chemical elements but also an indefinite series of indeterminate content and form. Metaphorically speaking, it is a breathing regime that reconciles the hostility between eagles and sheep – predators and prey – embracing "a real essence of man before the animal differences of eagles and sheep" (Laruelle 2018, 10). Amid the mechanized air of the Post-Anthropocene, non-air unfolds a void – a vacuum that does not negate fanaticisms but affirms them; a space that does not resist the trauma of internalization but facilitates the encounter with it, recognizing trauma and destruction as inherent components of air itself. With non-air, air draws near the feminine rûah. As discussed in the analysis of the phallocentric definition of air, rûah denotes an instance characterized by "inclusivity toward others, which may extend even to the stranger and alien" (Johnson 2018, 153). However, a non-philosophical interpretation of rûah implies more than just the "breath of God," which inherently entails further biblical determination. Rather, the non-philosophical view of rûah makes air embody a radical inclusivity toward the other or the stranger, defying superficial classification as divine or worldly. This concept of air's *expansion* and simultaneous *emptying* resonates directly with Laruelle's conception of thought: "Non-philosophy has finally found in the Stranger its strategically most adequate concept of man, more precisely of the subject as existing beyond the real immanence that it moreover is in its ultimate cause" (Laruelle 2013, 142). Similarly, it can be inferred that non-air finds its adequate concept in the *vacuum*, where breathing seems "actually" impossible. With non-air, we arrive at a juncture where air transcends its status as a mechanically infused medium circulating within breathing machines and tubes attached to humans. When humans realize they can cease conventional breathing without suffocation, they instinctively take their first breaths in the non-air. This engenders a distinct aerial togetherness devoid of oppression or enforced coexistence; a second layer of air emerging above the tender cruelty of breathing machines, with the need to seek and capture it again and again. As Cioran articulates: "Poverty is not a transitory state: it coincides with the certainty that (...) you must struggle for even a breath, and conquer air itself, and hope, and sleep" (Cioran 2010, 173). It is noteworthy that embracing the struggle to breathe beyond prevailing regimes can serve as a stimulant in introducing a new aerial condition. The vacuum is not a realm of suffocation but offers a space where new breathing possibilities start to emerge.

Notes

1 Han has described the processes of optimization and surrendering of subjectivity in an impressive way: "Life is reduced to a biological process that must be optimized. It loses any *meta-physical* dimension. Self-tracking acquires the status of a cult. Digital hypochondria, constant self-measurement with the aid of health and fitness apps, degrades life into a mere *function*. Life is divested of any narrative that could give it meaning. Life is no longer a matter of what can be *recounted* but a matter of what can be *counted, measured*. Life becomes bare, even obscene. There is nothing that promises to endure. All those symbols, narratives, and rituals that made life something more than mere survival have faded into oblivion" (Han 2021, 16).

2 Juliana Spahr's thought must also be viewed critically in this context: "How connected we are with everyone. The space of everyone that has just been inside of everyone mixing inside of everyone with nitrogen and oxygen and water vapor and argon and carbon dioxide and suspended dust spores and bacteria mixing inside of everyone with sulfur and sulfuric acid and titanium and nickel and min-

ute silicon particles from pulverized glass and concrete. How lovely and how doomed this connection of everyone with lungs" (Spahr 2005, 9–10). In Levinas' understanding, the ethical function of the lungs can and must never be understood as a "lovely" or "doomed," since such attributions are already based on established concepts that exclude the immediate entry of the other. Rather, Spahr shows a refined form of the soft language of bliss.

3 Levinas has rightly been characterized as the philosopher of the ethical – his position that ethics must be the first philosophy is well known – who gives the instance of the other the central role in his work. In his first major work *Totality and Infinity*, published in 1961, Levinas introduces the significance of the other: "The Other remains infinitely transcendent, infinitely foreign; his face in which his epiphany is produced and which appeals to me breaks with the world that can be common to us" (Levinas 1969, 194).

4 If we refrain from the literal association of catastrophe as "annihilation" or "destruction" and instead use the original meaning of the word – "reversal" – then this catastrophe suggests nothing less than the reversal of any human existence.

Part II

Matrixial Breath – The Rhythms of Feel-Knowing Aerial

4 Never Was Heaven

In general understanding, a vacuum is characterized by its sheer emptiness, symbolizing the absence of any life. Within the traditional Western paradigm of thought, reconciling emptiness with vitality appears implausible, as emptiness typically embodies a consistently negative principle from which nothing positive would emerge. Existing within a vacuum seems incongruous with the present mechanisms of reality apathy, given that apathy *presupposes* a precise technology – an array of apparatuses far from being empty and indeterminate. With their highly digitalized and globally dispersed processes, anesthetic instruments strive to prevent the vacuum – the condition of non-air – at any price, as it signifies nothing less than the condition for the inception of a new breathing dimension beyond anesthesia and lifelessness. Let us gradually explore the potentiality within such a vacuum.

*

For Cioran, the concept of emptiness does not denote negativity but, conversely, a state of fullness: a "seat of truth," "fulfillment," and "liberation" (Cioran 1968, 63). For those who enter such a condition,

> everything is offered to you, everything belongs to you; dominating space, you pass from one object to the next, from one world to the next. The universe at your feet, you accede at once to the essence of happiness: exaltation at the zero point of Being, life transposed, promoted to the state of breath, of eternity, which breathes and which no mystery burdens.
>
> (Cioran 1968, 203–204)

As highlighted by Cioran, such emptiness cannot merely be construed as a negation of air or vitality but rather as the *foundational condition* for eternal and liberated respiration – uninhibited by the mechanisms of phallocentric control.[1] Within such a vacuum, in a temporal and spatial domain of the "zero point" of humans' existence, a *yet unknown substance* emerges, transcending prevailing mechanisms and ideologies. Non-air enables humans to respire where they initially reach back to the origin of their being-human.

DOI: 10.4324/9781003603672-7

Cioran also explores the liberating essence of such emptiness when he posits that "freedom can be manifested only in the *void* of beliefs, in the absence of axioms (...) Freedom, I was saying, demands, in order to manifest itself, a *vacuum*" (Cioran 1987, 12). According to Cioran, all freedom, all liberated humanity necessitates emptiness, absence, and non-air, which, as Laruelle has shown, transcends the dualisms of the positive and the negative: an aerial state where life-giving and life-suffocating elements, trauma, and bliss perpetually overlay, without fusion or separation. This underscores the resistance of prevailing anesthetic mechanics toward the condition of non-air: it cannot be understood within established phallocentric frameworks, given its fluid, co-emerging nature, a condition of being both one *and* the other, transcending dichotomies of power. An empty framework of thought, where death and life, inside and outside, are equally justified and integrally interwoven, must be rejected by prevailing unifying forces due to its unpredictable dynamics, inclusivity, and indefiniteness: it actualizes nothing less than a virtuality, latent yet unacknowledged. François Laruelle contends: "This is a world that is immense as well as invisible, intangible, unobjectifiable, but which is perfectly conceivable once it is thought and the unitary hallucination has been dispelled" (Laruelle 2018, 18). Thus, approaching this void framework of thought entails independently unfolding one's previously unseen (or kept unseen) world and integrating it into one's life.

*

The counter-hegemonic and anti-phallocentric potential of such a void – the alternative to the mechanics of contemporary anesthesia – can clearly be discerned in the opening paragraphs of Cioran's *A Short History of Decay*:

> In itself, every idea is neutral, or should be; but man animates ideas, projects his flames and flaws into them; impure, transformed into beliefs, ideas take their place in time, take shape as events: the trajectory is complete, from logic to epilepsy... whence the birth of ideologies, doctrines, deadly games.
>
> (Cioran 2010, 3)

Once emptiness and neutrality are abolished, once breath is animated, once an aerial space is laid out where one particle stands above the other, the supremacy of ideology takes hold and neutral openness of all particular multiplicities is abolished and hierarchized: non-air transforms into dogmatized matter, into a controlled event to which all who breathe must submit. Power and barbarism take hold. Such indoctrination manifests as an operation of phallocentrism, the transformation of the indeterminate rûah by the masculine power of unity and totality. Entering the void, thus, not only represents a counter-hegemonic practice but also points to its inherent femininity. Such a connection between emptiness and femininity is recognizable in Cioran's pessimism, albeit only on closer view:

> Women are safe on the spiritual plane because in them the dualism between life and the spirit is less intense than in men. ... Women are ciphers: the more you think about them, the less you understand them. In front of women one is *silent* just as one is silent in contemplation of the world's secret essence. But where the latter is unfathomable infinity, the former is simple mystery, in other words, *void*. Not greatly disjoined from life, woman is a temporary salvation for those on the heights of despair, because through her a return to life's unconscious and innocent pleasure is still possible. Grace, if it has not saved the world, has saved its women.
>
> (Cioran 1992, 60)

This quote cannot truly illuminate Cioran's view of women; what seems essential, however, is that on the meta-level of the feminine, Cioran identifies a silence, wholeness, and void – in an immediately positive sense – that closely resembles the qualities of the feminine rûah.[2] The emptiness of the non-air thus correlates not only with "a" fullness but with the fullness of the feminine. If rûah manifests itself in this way – as a space of non-blissful breath, as a dimension of freedom – this has nothing in common with the absence of aliveness but only with the absence of anesthesia – calling phallocentric resistance to the scene.

*

The liberating emptiness of the non-air is not only evident on a conceptual level but also manifests itself in the spaces and landscapes of contemporary planetary trauma. What unfolds beyond the tender cruelty of an automated breathing regime and a blissful atmosphere of anesthesia – beyond the phallocentric mechanisms of oppression of the other – is a biosphere of emptiness: Landscapes of loneliness, of destroyed austerity, a habitat of "scorched earth" on which only vague traces can be made out; a non-place, an infinite barren land. Its sea only reflects the fire of the illusionary and ruptured atmosphere of internalization. Never was heaven, as the poet Paul Celan postulated in his poem "Eastersmoke": "Never was heaven. But sea still is, fire red, sea" (Celan 1995, 205). The emergence of non-air can be understood as the final act of a global requiem, which has extended over the "heavy plow," Boyle's "air pump," the mystery of insulation, the flattened breathing within the cover, up to the digitalized surrender of human liveliness – the dramatic closing scene in which the burning breath-constructions fall into a sea and onto an earth of emptiness and destruction, like inverted heavenly manna. But, as Celan expresses, this desolation *exists*, despite everything: "but sea still *is*." Despite all the drama, the spirit of freedom resonates in this post-apocalyptic scene, as the numbing atmosphere has ceased to exist in its internalizing and synchronizing manner. Indeed, this process of emptying is not a dying light but an eternal light: *Lux aeterna*, a positive ontology of breath beyond phallocentric control and suppression. Ultimately, light prevails. It could even be posited that this apocalyptic *lux aeterna* represents the quasi-meteorological condition for allowing the feminine rûah to blow (again, or perhaps for the first time?) and opening up the space for entering a liberating emptiness. Alenka Zupančič intriguingly noted that

> the crack in this looming fate does not imply that there is a chance, a possibility, or hope that it won't happen; it signifies that what is bound to happen (and will necessarily happen) is not fully there yet: this, and nothing else. ... The call to awakening potentially heard with the appearance of the Bomb is not simply "let's do everything in our power to prevent it before it's too late," but rather "let's first build this totality (unity, community, freedom) that we are about to lose through the Bomb." If through the Bomb "the idea of totality arises visibly and for the first time on our horizon" (as the idea that we could "lose it all"), then perhaps we have to take it from here and build on this.
>
> (Zupančič 2018, 22)

Lux aeterna does not signify, as in Béla Tarr's epochal cinematic apocalypse *The Turin Horse* (Tarr 2012), a darkening progression toward the end, but rather the recognition that, in the very final moments before the protagonists are engulfed in absolute darkness, a new, distinct, unknown, possibly eternal vitality emerges: "We can hear them breathing, only their breathing can we hear. Dead silence outside, the storm is over" (Tarr 2012). Especially in this bleak condition, *a new hopeful space begins to unfold.* Anxiety has fallen silent; the empty space has been opened: dead silence, igniting a spark of life. Despite everything, a light exists; not a light of doctrine, but a light of infinity.

*

Paul Celan shares a similar perspective, still more focused on language. In the so-called "Bremen Speech" of 1958, he speaks of "passing through" the inexpressible, which makes it possible to find language *in the first place*; he discusses the *necessity* of passing through the horrors of the inhuman, of entering this almost lifeless void to reach true humanity:

> Reachable, near and not lost, there remained in the midst of the losses this one thing: language. It, the language, remained, not lost, yes in spite of everything. But it had to pass through the thousand darknesses of death-bringing speech. It passed through and gave back no words for that which happened; yet it passed through this happening. Passed through and could come to light again, "enriched" by all this. In this

> language I have sought, during those years and the years since then, to write poems: so as to speak, to orient myself, to find out where I was and where I was meant to go, to sketch out reality for myself.
>
> (Celan 2001, 395–396)

Lux aeterna virtually requires passing through the inexpressible, breathing in almost absolute darkness, moving within spheres of non-air. But in this passing through, in embracing what could potentially kill oneself – nothing less than the fullness of true humanity emerges. At this point, it seems crucial to seize the burning pieces of the internalizing atmosphere and use their light to illuminate a yet unknown, different commonality and vitality. Paradoxically, the current planetary trauma, the arrival at a point where almost every potentiality of human life is about to be extinguished, signifies the emergence of life itself, in its simplicity and beauty: the realization that when "never was heaven," there is an opportunity to build what can come very close to such heaven; a practice of picking up and carrying along the splinters of an old world seems necessary; a practice that confronts the void to possibly unearth life's "*Urphänomene* [originary phenomenons]," as Hannah Arendt put it in regard to Walter Benjamin (Arendt in Benjamin 2007, 51). A paradoxical hope can be identified in the early 21st century, where all breathing seems impossible due to the anesthesia and digitalized suffocation. But perhaps it is precisely the current vacuum that, as a momentary window of opportunity, allows us to breathe for the very first time.

*

When Israeli psychoanalyst, philosopher, and artist Bracha L. Ettinger conceives of the contemporary human subject as a "subject after the death of the subject" (Ettinger 2015a, 343), then this observation assumes fundamental significance for the formulation of a practice of breathing non-air; does this observation not also imply that humanity emerges anew after the end of the human, or perhaps *must await* the demise of the former *conditio humana* to attain its genuine, immediate humane essence? In a more recent statement, Ettinger underscores the irrevocable nature of such a profoundly paradoxical transformation: "After the catastrophe, the humane must reappear – in wit(h)nessing" (Ettinger 2015b, 356). Humanity necessitates the antecedent catastrophe to foster its emergence; it requires exposure to lethal atmospheres to facilitate the emergence of salutary meteorologies. But how can breathing be delineated within such novel, divergent meteorologies, which commence amid the desolation of scorched earth, the void of the vacuum, and the recognition that "never was heaven"? How can we re-articulate respiration after the end of breath? To what degree does the enigmatic notion of "wit(h)nessing" contribute to addressing these inquiries? To provide an initial response, it becomes imperative to expound Ettinger's thought in greater depth.

*

Drawing upon Freudian psychoanalytic theory and its reinterpretation by Lacan, Ettinger has advocated since the 1980s for a theory of subjectivity that eschews oedipal reductionism and the repression of the feminine, instead positing a distinct configuration of human existence. This configuration is characterized, in particular, by an engagement with and a carrying of the trauma of the other – by an awareness that embracing catastrophe and weaving threads of care amid the traces of past cruelty are indispensable for guiding humanity beyond the realms of barbarism, oppression, and anesthesia. By stating that "we have to introduce non-phallic symbolic spheres alongside the phallic sphere" (Ettinger 1992, 178), Ettinger introduced an alternative view of human subjectivity and its agency: "I call one such sphere the Matrix. The Matrix conceptualizes non-one-ness, prenatal experiences of *I and non-I(s)* in co-existence without assimilation and without rejection" (178). What becomes clear with this initially abstract definition, which refers to the prenatal experiences of the not yet born child in the mother's womb, is a view of the human corporeality as constituted in a profoundly relational field. In these early prenatal stages of subjectivity, there is an immediate, fluid exchange between mother and unborn child that can be understood neither as total symbiosis nor total separation; in fact, any totality is alien to the matrixial condition, which sees itself constituted by multidimensional rhythms of "non-one-ness." These rhythms between mother and unborn child defy established psychoanalytic or psychological schemes; instead, they follow subtle and non-taxonomized communication rhythms on physical and psychic levels. The matrixial communication is an inter- and intra-active exchange of traces and fragments, an assumption of responsibility for and, equally, a traumatization of the other. Traumatization in a matrixial sense is a much more interrelated, mutually carrying action than the established notion of trauma as split set by the anxiety-inducing mother in the oedipal triangle. With Ettinger, trauma is "related to the truth of the relations between body and transcendence in the space-time between birth with an-other and the solitude in life, the solitude of death" (Ettinger 2019, 213). In this process of mutual traumatization, a co-lapse of gazes takes place between the mother and the unborn child. In contrast to the hierarchy of the phallic gaze of the split with its "impossibility of sharing trauma and fantasy," the co-lapsing gaze states that "there is an *impossibility of not sharing them*" (Ettinger 2001b, 132). The becoming subjectivity in the womb, but also in its extrauterine becoming as a young, adult, middle-age, and old-age human being, sees its body-psyche continuously overlaid and traumatized by human and more-than-human traces of the other: it co-lapses continuously. With the co-lapsing gaze, the reconsideration of the unconscious as an exclusive interior takes place, raising the concept of a *shared unconscious sphere* between mother and child, between the human and the more-than-human, or between the present and the past. Subjectivity based on such shared unconscious sphere is always "a singular shared trans-subjectivity" (Ettinger 2005, 704), where the notion of subjectivity shifts from its solipsist and oedipal understanding toward "a transgressive encounter between 'I' (as partial-subject) and uncognized yet intimate 'non-I' (as partial-subject or partial-object)" (703). Such a non-I entails the human and the more-than-human or even the unknown other. Trans-subjectivity, emerging in co-lapsing procedures of gazing-with, means a continuously forming aliveness. The human and the more-than-human as partial multiplicities continuously inscribe themselves into one another and relate to, care for, and exist with one another. As such, trans-subjectivity abolishes the hegemonic notion of an autonomous or pre-defined self in favor of a "body-psyche-time-space," a transitive "web of several" (Ettinger 2005, 704). This web of subjectivity appears less as the Freudian three-story topography and bears no resemblance to a controlled atmosphere of bliss. Rather, such a web can be understood in aerial terms as a "breathing house," the constitution of which was aptly described by the poet Rose Ausländer: "Invisible bridges span / from you to humans and things / from the air to your breath / speaking with flowers / as with humans / whom you love / living in the breathing house / a human flower time" (Ausländer 1987, 42, translation mine). In this perpetual inscription, which takes place in the Ettingerian matrix, a body – "a house" – stretches out on threads or even breaths, which overlap with one another, merge into one another, and "span invisible bridges," forming an ephemeral, shared constitution. "Gazing-with," the mutual gazing into each other, can thus be understood as a "breathing-with." Significantly, in such a constitution of human subjectivity, what was previously considered outside the body-psyche loses its otherness and actualizes itself as part of my I, as non-I.

*

The mutual inscription into such a "breathing house" unfolds the concept of a matrixial love, which has little in common with traditional concepts of economy or reproduction but is to be understood as fundamentally transitive: "A matrixial love is care-full and com-passionate, yet painful because of the inevitable processing of the other's trauma and because of the inevitable participation in the transformation and opening of boundaries for transmission or reception, fragmentation, contracting and withdrawal" (Ettinger 2010, 205). Ettinger proposes an ephemeral form of love based on mutual resonance and awareness of the archaic jointness within the womb. Matrixial love embraces the dynamics of "carriance" (Ettinger 2015b, 353) and emerges on its very proto-ethical ground: mutual carrying is an invitation, a gift, but it is by no means guaranteed that this gift will be accepted by oneself or the other. Matrixial love does not entail the mere acknowledgment of the other, but a fundamental *trust* in the other and "feel-knowing the strings trembling between subject and subjects or objects or things" (Ettinger 2015a, 345) in severality. A different consciousness and constitution of subjectivity emerge in the awareness and affection of these relational strings and resonant connections. Matrixial love as life-determining agency means an attentive, careful trust and the affirmation of the principle that "*I am* thence *I was carried*. I am therefore *I will carry*" (344). It suggests the conscious introduction of an ethical paradigm of "coemerging, coinsiding, kernel to kernel knowing in one another" (344), the resistance to endless fragmentation, and the negation of exclusive separation and solitary objectivity. With this, Ettinger introduced a psycho-physical relationship, which suggests a loving connection that "occurs not only by *membranous folds* but through transitive mental vibrating strings" (347). Here, it becomes evident that every human individual is fundamentally one who *was, is, and will be a possibly carrying and carried one*, and as such, participates in ceaselessly *possible* copoietic processes: the flattened "we" of subjectivity turns into severality that depends on affect. Every moment of life becomes a choice and a "passage from response-ability to responsibility, from wit(h)nessing to witnessing, recognizing the difference between non-life and death and the passage from non-life to life" (343). The paradigm of matrixial love and the agency of the "maternal-matrixial eros" (Ettinger 2014, 124) suggests the affirmative inclusion of the other and the other's inexpressible real, her/his/its trauma and wounds, only to realize that the scars of the other are also my own. Such awareness is a highly virtual one, located in a spatial dimension that Ettinger calls "subreal": "Subreality, which admits a kind of 'quantum-like' resonance of I/non-I elements, depends on our capacity to self-fragilize – to reach some awareness of that level of transconnectedness between different elements in between me and the other, me and the cosmos, beyond the threshold of our sense-perception" (Kaiser and Thiele 2018, 117). In Ettinger's case, matrixial love does not mean an identarian, representable love, but the unrepresentable recognition of the other in oneself and one's ethical obligation to carry the traces of the other's trauma further, as well as one's trauma being carried by the other. In matrixial love, identities are fragile, open, and porous, manifesting a subreal intertwining beyond symbiotic mechanisms. It includes the knotting of "subreal strings" (Ettinger 2019, 208) across subjects, objects, places, and times, and the affection for the "vibrations arriving from inside herself and from the inside of the outside" (Ettinger 2015a, 347). It is a love of a "heartbeat (…) off-time" that "initiates combat against repetition of senseless cruelty" (343).

*

The carrying respiration within this resonating "breathing house" finds articulation in the Ettingerian concept of "wit(h)nessing" (Ettinger 2015b, 361), wherein Ettinger claims that "wit(h)nessing is becoming a participatory witness" (Ettinger 1999b, 94, footnote 12) or, with greater precision, "wit(h)nessing is becoming a participatory witness to an event one did not directly experience" (Ettinger 2020, 399). Grounded in matrixial love, wit(h)nessing unveils a paradoxical togetherness: the perception of unknown traces with which the respiring human appears to share no commonality.

Simultaneously, however, through their respiratory perception, the human recognizes that these enigmatic signs materializing in the air, on the earth, in the face of the other *nevertheless* exert an influence upon them – without a clear understanding of why. In weaving "subreal strings," the human intertwines with these traces without comprehending their origin, location, or essence. The one who thus weaves in such matrixial way, who embeds in the modalities of matrixial love and matrixial respiration, embodies a fundamentally unknowing being. The only dimension evoked in such relationality is *concern*. In being concerned, the duality of interiority and exteriority, previously subject to instrumentalization and manipulation in diverse forms, dissolves. In wit(h)nessing, it becomes evident that the totality of interiority and exteriority can no longer be justified, as it reveals that the otherness of the scorched earth, the collapsing atmosphere, the fellow human beings cannot, and must not remain foreign – but rather begin to concern me, demanding shared suffering and assumption of responsibility. The other ceases to be the total other but aligns within "a singular shared trans-subjectivity" (Ettinger 2005, 704), wherein self and other continuously overlap, inscribe themselves into each other, and *respire towards each other*. Such mutually affected respiration no longer transpires within a transcendent exterior space or a controlled interior space extending into the mechanisms of breathing machines – but within this other, alternative dimension of concern, within this expansion of one's lungs amidst a (post-)apocalyptic void. The inscription of the other into one's corporeality ruptures any totality. In the matrixial trans-scription, a yet unnoticed, yet rhythmically primordial pulsation of respiration unfolds – "care-full and com-passionate, yet painful" (Ettinger 2010, 205), or, as Cioran frames it: "exaltation at the zero point of Being" (Cioran 1968, 203–204). This is the pulse of the matrixial. Respiration transforms from a mere biological function, or a linear attachment to a breathing mechanism, into an ethical function of concern. It is no longer defined by an atmosphere – an aerial sphere, whether open or enclosed, free or controlled – but by a *breathing desolation*, by inhaling non-air, which for the first time elucidates that respiration beyond prevailing biological paradigms of uniformity is *conceivable*: breathing within traces, breathing within perception, breathing within participation, in carrying. Such a moment of realization might bring about horror and trauma; it signifies the dawn of a different world. The human experiences a bodily function and responsibility yet unexplored. An unprecedented dimension intrudes here, previously precluded from respiration. The human recognizes that in breathing within this barren land, amid horror and trauma, no "horror of death" (Ettinger 2011, 3) prevails, contrary to all assumptions and apprehensions. Rather, the uncanny [unheimlich] transmutes into a condition of "heimlich," further characterized by Ettinger in *The Matrixial Borderspace*:

> Erotic aerials weave together partial-subjects and -objects that unknowingly, and unknown to each other, are caught up in a process of co-in/

> habit(u)ating. In this trembling *heimlich* effect, experience and meaning emerge. A rhythm of swerving in relation, conducting toward an unforeseen home *it itself* is creating, bypasses the off/on beat. This rhythm never arcs as far as a total "on" or "off," for it treasures several celibate spasms at the same time and is attuned by several co-spasmings. Co-in/habit(u)ation is primarily affective and resonant (...) A center-less *heimlich* affect silently ascends behind the "unheimlich" aesthetic experience (Freud), anguishing and soothing, tearing and stitching the wounds of a nomadic place, opening in it a rhythm of interval for an exile, suspended like a rotating sea-wave between its fading and a next birth.
>
> (Ettinger 2006a, 159–60)

Through the respiration of concern, a new rhythm emerges leading to exile, to this yet unknown "breathing house," whose dimensions cannot be delineated by binary poles of "on" and "off," but rather constitute a domain of "co-in/habit(u)ation," transcending mere materiality and characterized by superimposed strata, drifts of air, and trauma particles: a breath of poetics, beyond deterministic biology. An entry into the darkness of language, an embrace of the ominous silence, a pulmonary movement akin to a "counter-rhythmic rupture or 'cesura' that affirms the freedom of an extreme moment in the neighborhood of silence and the breakdown of language, insanity, or death. Poetry is the spasm or black-out [syncope] of language" (Clark 1997, 260). As Nieuwenhuis (2015) observes, such poetic respiration manifests in its disruption and departure from dominant epistemological modalities, preceding them and resisting any subsequent attempts at assimilation (Cf. 178). Conscious that "never was heaven," humanity starts to respire anew: respiration that remembers, considered impossible until now. Humanity becomes alive.

Notes

1 Cioran's notion of emptiness and Heidegger's concept of nothingness must be differentiated here. In Heidegger, a linear temporality is requisite for comprehending "Dasein," and human existence must be within a clearly delineated temporal horizon. For Heidegger, this horizon is death – the end of Dasein – which asserts that "Da-sein reaches its wholeness in death" (Heidegger 1996, 221). Heidegger's "being-in-the-world" can thus be fundamentally construed as a "being-toward-death" [Sein zum Tode] (233). This state of existence approaching death is intimately entwined with the notion of anxiety: "That about which one has anxiety is being-in-the-world as such" (186), implying that anxiety also invariably gestures toward death. The fear of death, for Heidegger, emanates from the fact that death always retains an indeterminacy – a nothingness and nowhere – which persistently eludes clarity: "The fact that what is threatening is nowhere characterizes what anxiety is about. Anxiety 'does not know' what it is about which it is anxious (...) Therefore, what is threatening cannot approach from a definite direction within nearness, it is already 'there' – and yet nowhere. It is so near that it is oppres-

sive and stifles one's breath – and yet it is nowhere" (186). In his 1929 lecture "What is Metaphysics?" Heidegger delineates such nothingness, which fundamentally sustains anxiety's foundation and the phenomenology of existence, more precisely: "For the *nothing is the negation* of the totality of beings [Allheit des Seienden]; it is nonbeing pure and simple" (Heidegger 1998, 85). The prevailing assumption that nothingness inherently signifies a negation of being and precludes vitality must thus be attributed to the Heideggerian argument, yet, as I argue, it is not universally applicable. Indeed, this book seeks to transcend Heidegger's prevalent phenomenological and existential legacy, as "being-toward-death" precisely embodies a fundamentally phallocentric conception, echoed, for instance, in Freud's notion of the death drive. The notion that nothingness excludes vitality and, therefore, authorizes anxiety could instead be construed as a necessary mechanism for controlling power structures, as exemplified in the mechanisms of breath control. However, if nothingness is construed as a *positive potentiality*, as in Cioran's thought, then such a notion can indeed be regarded as subversive. While Cioran's emptiness and Heidegger's nothingness may employ analogous terminology, upon closer look, they are fundamentally incompatible with each other.

2 The biblical book of Isaiah explicitly speaks of a silence that is characteristic of the rûah: "Here is my servant, whom I uphold, my chosen one in whom I delight; I will put my Spirit on him, and he will bring justice to the nations. *He will not shout or cry out, or raise his voice in the streets*" (Is 42:1–2).

5 Fragile Trans-Subjectivity

If breathing emerges as a carrying phenomenon of concern, reciprocal inscription, compassionate traversal of darkness and trauma, and co-in-habit(u)ation beyond the deterministic biological function of respiration – then such a perspective is not merely a zeitgeist-driven expression grounding in the presence of feminist discourse in contemporary socio-political realms.[1] Instead, it is necessary to speculate on a deeper stratum of significance for such an alternative respiratory paradigm. At this juncture, let us etymologically trace the imperative of the human subject *as one who carries and is carried*: Recognizing an etymological relation between the words of "subject" and "carry" in Hebrew, Bracha Ettinger stated that "while the Latin term *subjectus* is linked (...) to 'subject of attributes and predicates' and to subjugation – this is very different in Hebrew where (...) 'subject' does not indicate being subjugated, but on the contrary: CARRY! Bear! Support!" (Kaiser and Thiele 2018, 110). Ettinger understands subjectivity as profoundly relational, in that each subject in its constitution refers back to the first prenatal encounter with the carrying mother in the womb: "each one of us, who came to life, came to life *since she was carried* inside a female body for a long-enough duration" (123). The traces of this archaic "encounter-event" (104) are carried in and with each subject throughout her/his life. Such traces, Ettinger noted, are traces of carrying: both in a literal sense – as when we say a child is being carried – and metaphorically in the sense of holding, inviting, or giving support to the other. The notion of carrying suggests compassionate and caring carrying: it is "care-carrying," an expression of "carriance" (104). With this carrying inscription in the constitution of subjectivity, the Ettingerian matrixial theory formulates an important alternative to the phallocentric tradition of thought. Only when we are carrying, in the sense of "carriance," we are true subjects: "I care-carry *ergo sum*" (107). However, it is not only through this feminine approach that any subject exists as a carrying being but also through the double meaning of the Hebrew root N.Sh.A, which means both "subject" and "carry": "the subject is NOSSE [נושא] (again from the root N.Sh.A), meaning: 'to carry'" (104). This inseparable link between "subject" and "carrier" relates to a significant ethical insight reported at the beginning of the Old Testament: After Adam and Eve were expelled from

DOI: 10.4324/9781003603672-8

paradise by God, Eve gave birth to two sons: the first she named Cain and the second Abel. As soon as they were grown up, both sons engaged in work: "And Abel was a keeper of sheep, but Cain was a tiller of the ground" (Gen 4:2). As it was common at the time, both sons regularly brought offerings to God. In the scriptures, a particular incident is now recorded: "And in process of time it came to pass, that Cain brought of the fruit of the ground an offering unto the Lord. And Abel, he also brought of the firstlings of his flock and of the fat thereof" (Gen 4:3–4a). What happens right after this offering cannot be fully understood. Indeed, God saw both offerings – Cain's fruits and Abel's meat – but he accepted only one: "And the Lord had respect unto Abel and to his offering: But unto Cain and to his offering he had not respect" (Gen 4:4b–5) – why this happened we do not know (is it already a rejection of Cain's pre-anthropocenic interiorizing practice of doing violence to the ground?), the result, however, is well known. Cain became envious of his brother Abel, with his body-psyche turning disappointed and sad: "And Cain was very wroth, and his countenance fell" (Gen 4:5). As a response to Cain's distress, we now hear a first voice, spoken directly from God: "Why are you angry, and why has your face fallen? If thou doest well, shall it not be lifted up? and if thou doest not well, sin lieth at the door" (Gen 4:6–7). In a recent essay, Ettinger shared an interesting etymological observation. For the phrase "shall it not be lifted up" – the consequence of "doing right" – the Hebrew Bible uses the word ShETH [שאת]. Ettinger notes that "the literal and figurative meaning of ShETH [שאת] is to bear, to tolerate, to lift, to suffer (...) I've checked many translations of this phrase and in all of them, we more or less find that ShETH [שאת] is understood as lift, as in lift your head high" (Kaiser & Thiele 2018, 112–113). However, the translation that if you do right, you can lift your head high irritates, as it does not imply any ethical obligation: Is it enough to lift your head if you want to do good? Ettinger suggests that the word ShETH indeed has ethical implications, "but this kind of suggestion is omitted from the common interpretations and translations of this passage" (112). The connection "if you do right" with "you may look up" does not seem to define an ethical position; the linking of the two propositional variables seems to fall flat. However, this is not the case when an alternative translation based on the etymological roots of the word ShETH is applied. Again, Ettinger:

> The literal and figurative meaning of ShETH is to bear, to tolerate, to lift, to suffer. But to begin with: *to carry*. The grammatical time of the sentence in Hebrew is future tense, but a kind of future, which in biblical language signifies also the past, that is, a verb that intends both past and future. "If you will be doing what is good: carry." (...) Carry – in the past and in the future. Then the sentence continues: And if you don't do what is good – so, if you don't carry your brother, your fellow human being – sin is crouching at the door: temptation of rage,

> violence (...) the desire to abandon and betray, the desire to possess and appropriate the other, not-carry.
>
> (113)

*

This new interpretation of God's response reveals a true ethical obligation: *if you want to do right, carry!* Not: if you want to do right, lift your head high. The existence of an ethical attitude is measured by the extent to which I carry the other. In the beginning, there is "carriance" (103), which is more than care:

> a kind of engagement with the other and the world, even when one is alone, even without relations, even when the other is no more, even when the world has gone, even from before we are humane. Carriance is beyond relationality; it is nonreactive, unconditional, and yet responsive, communicaring (...) Not just (...) before my face – but (...) from the before of the in-front-of-my-face, which I understand as an indication of a kind of invisible source that is symbolizable, a time before our human time, literally signifying even from the before of my facing something. The time of the sacred perhaps, the sacred in and for the human – as that which is before the human and in front of her face.
>
> (115)

Referring to the Old Testament incident, Ettinger points out the essential paradigm of being human in carrying the other in all her/his/its historical complexity. A paradigm of carrying the wounds and traumas of the other; of carrying the times of the other, even if they reach far back into the past; of carrying the voices of the other, even if they are not perceivable anymore.

*

Once again, we find an etymological nexus, as ShETH [שאת] shares roots with NOSSE [נושא]. When Ettinger asserts that the subject must always be comprehended as a carrier (NOSSE), while simultaneously positing such carrying (ShETH) as the primal ethical command of God upon humans, it implies that not only is the human a carrier but also that humanity can only be fundamentally achieved when one starts to carry, a practice for which the human must continually decide anew. To be truly human is to carry – to rhythmize one's relationship with the world, one's inhalation and exhalation, one's respiratory mechanism in a manner determined by an essential concern for the other and a readiness to carry this other. This readiness, rooted in its intrinsic feminine matrixiality, is constitutive: "Whoever takes upon itself the symbolic ShETH [שאת] enters a feminine-maternal-matrixial zone of ethics that is beyond gender and social function. To carry is the primal mode upon which responsibility for the other is patterned – or rather, I prefer – *mattered*" (Kaiser and Thiele 2018, 116). Such carrying impulse of breath unfolds dimensions of humanity, already inscribed within individuals through their matrixial becoming:

> Humanizing means turning such a potential engrained in our real into a symbolic zone in which the destiny of ethics is appraised. Co-inhabit(u)ation in the humane and the linkage to the Other under this feminine function is perhaps the divine in the human to which we can return as a metaphor: *I must carry you* can be a crucial intervention of the feminine. It has its price though, and each time a different one.
>
> (Kaiser and Thiele 2018, 122)

The call for concerned and carrying breathing is invariably coupled with a *decision* that the human must continuously make with each inhalation and exhalation: the readiness to render oneself fragile and receptive to the other as the (non-)I. Historically, this decision has been predominantly made one-sided in Western philosophical thought, marked by anthropocentrism and individual unity. Roland Barthes aptly delineates this history concerning the inviolability of the author in literature:

> The author is a modern figure, a product of our society, arising from English empiricism, French rationalism, and the personal faith of the Reformation, which discovered the prestige of the individual, or, more nobly put, the "human person" (...) The image of literature in ordinary culture is tyrannically centered on the author, his person, his life, his tastes, his passions.
>
> (Barthes 1977, 142–143)

This totalitarian authorial uniformity is now juxtaposed with a fragility understood in matrixial terms, as hinted at by Barthes:

> We now know that a text is not a linear string of words releasing a single "theological" meaning (the "message" of the Author-God) but a multidimensional space in which a variety of writings, none of them original, blend and clash. The text is a tissue of quotations drawn from the innumerable centers of culture.
>
> (Barthes 1977, 146)

Barthes' speculations into the nature of the text correspond to our understanding of carrying breath: the human breathes-carries and transcends individual prestige; her/his corporeality transforms into a "multidimensional space" where various particles of breath from others, carried through the air, interfere and affect one another. In this exchange, the human body becomes the medium for Ettingerian co-inhabit(u)ation, a body-space within the integrating breath house elucidated in the poems of Rose Ausländer; a permeable corporeality, relieved of the dominant position it held for centuries. Simultaneously, however, this suspension of individual human dominance does not entail a shift from the individual to the "dividual," as posited by Deleuze in his discussion on control regimes. Fragilization of the body does not imply standardization or objectification that renders the individual quantifiable and thus controllable. On the contrary, fragilization entails a movement toward the other, an acknowledgment that the self can only be constituted by the other and thus remains indeterminate. This entails relinquishing one's dominance and becoming vulnerable to the breath of the other. As Ettinger articulates: "The I and non-I are wit(h)nessing one another, and by that they become partialized, vulnerable, and fragilized" (Ettinger 2005, 704). In this mutual awareness, partialization, and fragilization, a breathing with/in/through/for the other emerges – a carrying breathing. I breathe with/in/through/for the other because the other is not alien but (non-)I, simultaneously distant and close to the self. The hegemonic unity of subjectivity undergoes a profound transformation into a *trans-subjectivity*, where the I is coupled to its (non-)Is, as Ettinger suggests in her formulation of "transconnection."[2] This concept invariably evokes archaic prenatal experiences within the womb:

> I, the one who came into life, have been transconnected to a female-body-psyche (non-I) with whom psychic strings, particles and imprints were shared and exchanged, entangled. To rethink the human subject as impregnated by an *I and non-I* transjectivity, and thus to recognize the human subject as nestling, co-inhabiting, co-inhabituating, co-emerging, and co-arising with-and-inside-and-outside of an other, is to recognize the importance of our net of strings to the structuring of each individual self.
>
> (Kaiser and Thiele 2018, 107–108)

From a literary standpoint, Roland Barthes underscores the Ettingerian phenomenon of "co-arising with-and-inside-and-outside of an other," by stating that

> a text is made of multiple writings, drawn from many cultures and entering into mutual relations of dialogue, parody, contestation, but there is one place where this multiplicity is focused and that place is the reader, not, as was hitherto said, the author. The reader is the space on which all the quotations that make up a writing are inscribed without any of them being lost; a text's unity lies not in its origin but in its destination. Yet this destination cannot any longer be personal: the reader is without history, biography, psychology; he is simply that someone

> who holds together in a single field all the traces by which the written text is constituted.
>
> (Barthes 1977, 148)

After the death of the author, a text's meaning rests with its recipient, who perceives herself/himself as an open, absorbent, and vulnerable container wherein particles, threads, and contextual meanings intertwine and complement each other. This signifies that humans never breathe solely for themselves but always in relation to others; we do not respire in isolation; rather, the vital exchange of oxygen relies on *the other individual* absorbing our breath, being affected by it, and transmitting it – much like how we absorb and carry the breath of the other. Thus, human life grounds on the other, rendering the self fragile, permeable, and fragmentary, realizing its existence within a sphere of (non-)Is, constituted and transformed in the pulsating matrixial exchange with these (non-)Is. Establishing a clear identity becomes untenable, as does mechanically embedding oneself in unitary relationalities. The I emerges when, and only when, it carries the other and is carried by it – when the other's darkness impacts it, when the I offers itself to shoulder that darkness, and, in turn, relies on being accompanied by the other. Ettinger observes that the process of "inscriptive exchange between/within several matrixial entities (…) dissolves the individual borderlines so that they become *thresholds* that allow a passage, which, for each participant, captivates what I call a surplus of fragility" (Ettinger 2016a, 156). This "surplus of fragility" ultimately catalyzes *metramorphosis*, which transforms the condition of a mutual, co-affective inscription of jointness-in-difference into a real, applicable, but fundamentally alternative form of ethics, politics, and theory of the subject. "The borderlines (…) are surpassed and transformed to become thresholds. When these transformations relate to transformations in the borderlines (…) metramorphosis may occur, creating redistribution in the shared field and a change in the common subjectivity" (Ettinger 1992, 200).

*

The subjectivity that absorbs the breaths of the (non-)I, carries it forth and imbues it with meaning via its concern can only be regarded as trans-subjectivity, as a corporeality transcending anthropocentric unity and dominance. Trans-subjectivity is wafted through trans-subjects and trans-jects, deciding with each breath to accept or reject these breaths of the other. Ettinger regards such practice as fundamentally rooted in the role of the aesthetically perceiving artist but extends it to anyone *deciding for* the carrying dimension of breathing and self-fragilization: "Actively-passive self-relinquishment in com-passionate hospitality begins with aesthetic and ethical desire and an actively-passive 'decision' of the artist to fragilize herself and loosen her psychic boundaries in order to surrender to vibrations arriving from inside herself and from the inside of the outside" (Ettinger 2015a, 347). It is crucial to note that these vibrations, traces, or breathing particles act on both trans-subjective and transjective levels: they may originate from other human subjects (trans-subjective level) or from other more-than-human entities (transjective level) – the non-I extends beyond the realm of human exclusivity. Ettinger clarifies that "matrixial trans(sub)jectivity with its vibrating strings that attune transjectivity (subject-object) and transubjectivity (subject-subject) in resonance, dissonance, and consonance" is "not limited to the maternal nor to pre-natality and pregnancy. They [these processes] occur at any moment, in any relationships, in men and in women, in-between, all throughout life" (Ettinger 2016b, 285). In another essay, Ettinger extends these processes explicitly to the more-than-human domain: "Consciousness is penetrated and enlarged through metramorphic reattunement with the other and with the cosmos" (Ettinger 2015a, 349), elaborating that

> it is reached by borderlinking one's own soul-psyche to the breath of the psyche of the other and to the spirit of the Cosmos. In each particular copoiesis with the Other or with the Cosmos, being-with and being-in is "self"-differentiation and individuation within transgressive reattunement, which creates spiralic metramorphic vectors.
>
> (Ettinger 2005, 708)

Resonance in carrying breathing thus occurs between subject and subject and between subject and object, suggesting that the (non-)I may originate from a more-than-human or even unknown milieu, as much as from the human realm. Such a matrixial copoiesis, characterized by the obligation to carry the breath of the other and the assurance of being carried by the other, refers back to the notion of matrixial love. Unlike symbiotic fusion, matrixial love embodies reciprocal care and an openness to the manifold breath drifts of others affecting the I. This concept also facilitates an exploration of the resonant breathing rhythms inherent in trans-subjective and transjective relations, surpassing mere symbiosis. A brief allusion to the theory of autopoiesis, where love similarly features, is needed at this point.

*

Humberto Maturana and Francisco Varela, along with Lynn Margulis and James Lovelock, are the most prominent advocates of an autopoietic view of planetary life, where the concept of love is key to any social system: "Among human beings, the basic stabilizing factor in the constitution of a social system is the phenomenon of *love*, the seeing of the other as a partner in some or all the dimensions of living" (Maturana and Varela 1980, xxvi). Maturana and Varela's concept of love means the love of one's own and the other's self as constituent parts of a planetary autopoietic view. Self and other contribute to establishing and maintaining a micro-mechanism essential for the functioning of the symbiotic and symbiogenetic procedures of the earth. Such a view of love suggests a secluded constitution of human subjectivity as a closed-off part of an incessantly operating system with a fundamentally egocentric perspective since loving the other eventually contributes to maintaining one's own being.[3] Both in Heideggerian ontology and in autopoiesis, it becomes clear that both "love" and "care" are organized by the relationship with and dependence on the other to bring forth one's autonomous world. This bringing forth, of which both Maturana and Varela[4] and Heidegger[5] speak, takes place in chains of dependence between the self and others, which are part of other chains of dependence within an all-embracing planetary symbiosis. The "concerned being-with with the others" (Heidegger 1996, 274) refers to an activity of creating individual, solitary worlds as parts of a vast global dynamic. In the correlation of Heideggerian care and the mechanisms of autopoietic love, a symbiotic planetary "environment (...) continuously regulated by life, for life" (Margulis and Sagan 1997, 94) emerges. At this point, the symbiotic and autopoietic view was now subjected to a fundamental criticism by Ettinger when she stated: "With Margulis' idea of symbiosis, what I find a difficulty does (...) concern (...) that her thought might lead to the denial, for good and for bad, of the option of carriance" (Ettinger 2015a, 347).

*

Ettinger's matrixial theory correlates with Leo Tolstoy's claim that "everything is, everything exists, only because I love" (Tolstoy 2010, 1058). However, contrary to autopoietic and symbiotic perspectives, Ettinger does not suggest the ossified love of oneself and the other to maintain one's existence, but a thoroughly *transitive* form of love. Beyond the mechanisms of autopoiesis and symbiosis, Ettinger proposes an ephemeral form of love based on mutual resonance, establishing a proto-ethical ground: a psychophysical relationship, which neither resembles *horizontal* symbiosis nor the human being as a *vertically* sexed individual who has not been transferred from the immortality of the prokaryotic symbiosis into temporal beings-toward-death, defensively remaining separated from the bacterial microcosm (Cf. Margulis and Sagan 1997, 95). If the subject wants to do justice to its obligation as a carrier and seeks to actualize this carrying ability through affected breathing with/in the other, then the Ettingerian concept of *copoiesis*, assumes significance. In copoiesis, matrixial love transcends the confines of individual being, with an "openness to the uncognized world and to unknown but intimate others by a compassionate Eros that is not a sexual libido in the usual sense" (Ettinger 2005, 709). Unlike autopoiesis, copoiesis embodies a relational dimension with the other devoid of utilizing them solely for self-maintenance. It exemplifies a genuinely human form of love characterized by "compassionate hospitality and generosity" (707). Here, the reception of the other transcends the dynamics of sexual libido, embracing the other in manifold ways. It is not a love oriented toward reproduction but rather a primary love of boundless perception, resonating deeply with the other, whether human or more-than-human. Matrixial eros does not foster a secluded form of subjectivity but cultivates a trans-subjectivity that flows through and carries one another. Such a transitive subjectivity arises from opening one's corporeality to the other and committing to engagement with the other. Genuine humanity can only emerge when humans open their body-psyche, relinquishing their centrality and opening them up to the other while embracing the imperative to carry this other and fulfill their human obligation *as carriers*. Such matrixial love is non-sacrificial, eschewing any symmetry: "To be witness to the invisible of carriance and to carry without sacrificing – it is for this that the maternal-matrixial as a form or structure of conscience needs words. Witness, not as a form or structure of that which is non-sacrificial, the beyond of the 'being-with-oneself.' Being-with-oneself with-in and in-resonance-with another" (Ettinger 2015b, 365). However, autopoiesis is inherently tied to an overarching symmetrical structure, operating through communal and sacrificial procedures. In autopoietic love, there exists no carrying without reciprocity, no acceptance of the other without preconceived consequences; the economy always persists: "Margulis's concept of sex is based on the symbiotic love of two Gaian cells ready to die to become one" (Anderlini-D'Onofrio 2009, 110) and further: "They are a web of interdependent cells whose symbiotic love flows like an endless, self-regenerative autopoiesis" (65). Autopoiesis contrasts the Ettingerian copoiesis, as it fails to recognize "compassionate hospitality and generosity" in its mechanistic biologistic frameworks; the other is merely perceived as a resource, with the otherness kept unseen.

*

Matrixial love, with its subreal relationships beyond autopoietic or symbiotic mechanisms, can be understood as an essential characteristic of the carrying breathing between the human and the more-than-human. This respiratory sharing is no economic appropriation of the more-than-human; neither is the more-than-human seen as a static extension of the human nor part of a mechanically understood planetary symbiosis. The matrixial theory considers the human subject as "a single shared trans-subjectivity" (Ettinger 2005, 704), where the notion of subjectivity shifts from its solipsist understanding toward "a transgressive encounter between 'I' (as partial-subject) and uncognized yet intimate 'non-I' (as partial-subject or partial-object)" (703). Such a radical re-constitution or rather *co-constitution* of subjectivity makes it clear that from the very beginning of being human, humans manifest themselves as a permeable fabric in which the traces and residues of manifold planetary affects and agencies have been inscribed and continue to inscribe themselves. This also means a significant change in the relationship between humans and nature. With the Ettingerian matrixial theory, the observation of a "bond between psyche and cosmos" (Roszak 2001, 290) becomes apparent. In this bond, it seems more appropriate not to speak of *the* other as a singular entity but to acknowledge the multitude of *others*, which *have always been* inherent parts of one's subjectivity. Since "the human spirit is a particle of the spirit of the cosmos to which we are borderlinked" (Ettinger 2015a, 350), a responsibility of recognizing, carrying, and caring for the traces of *my others within me* unfolds. This mutuality, however, is not about a thoroughly harmonious homeostatic balance but rather about a fragile subreal condition in which jouissance, joy, and hope are interwoven with the subject just as much as the trauma, horror, grief, anxiety, and unspeakability of the (post-) anthropocenic real. The proto-ethical obligation to absorb the breath of the other, to be affected by it, and to carry it further is encountered when traversing these real virtualities with one's own body and breath. Such traversing procedures allow actualizing "polyvocité (...), agencements collectifs (...) [and] systèmes ouverts"[6] (Guattari 1973, 260–261), which can lead to a fragile and communal "response-ability" (Ettinger 2005, 706) that enables narratives and practices of inclusive respiratory relationships, which are still unrecognized but which *have always been and still are* close to every human being. In this sense, otherness can be understood in a non-dichotomic sense, which does not deny or negate dichotomy but leads to a physical and psychic condition in which established dualisms – across space and time – can be *mutated* through "compassionate hospitality" and "cosmic consciousness" and pave a proto-ethical path that shows that "the Earth hurts, and we hurt with it" (Roszak 2001, 308). The recognition of the other in human subjectivity via the Ettingerian matrixial theory allows for thinking with and within the planetary others and advances toward narratives of a copoietic breathing co-existence beyond any synchronicity. The interwoven self suggests an identity that unfolds not as a closed entity but through the absorption of the others with every breath, thereby embodying the open and permeable bodily systems suggested by Guattari above. It acknowledges that every human is inherently a carrying body, capable of breathing, i.e., incorporating the others into the self and constituting one's vitality. Such a constitution invariably necessitates the involvement of at least two actors. Still, in reality, it involves a manifold number of human and more-than-human agencies whose breaths intersect, intertwine, and resonate in subreality. Each breath is a fragile resonance within this aerial dimension of others' breaths seeking absorption to continue living. Such an aerial space transcending enclosure (atmosphere) or flatness (cover) must be further delineated as a space enabling welcoming carriance: carrying breathing calls for a carrying airspace.

Notes

1 A discursive presence that only superficially reveals a subversive femininity as it appears in Ettinger's thinking, but which in its basic features remains permeated by phallocentric mechanisms such as the split. In an interview on the topic of

"On Sex Without Identity: Feminist Politics and Sexual Difference," published in the *Los Angeles Review of Books* in October 2019, Alenka Zupančič interestingly states: "So you need to make this framing appear within the picture itself and problematize it. And feminist struggle did this. When it started to occur as a political movement, it focused on inscribing the *split*, as a division between different worlds (public and private, masculine and feminine, et cetera), *into one the same world*. Society is not composed of man and women; it is *split*, and this split is repressed. This is not the same as to say that women are repressed. Women were, are, oppressed, but this is not the same as repression, in the psychoanalytical sense of *Verdrängung*, of the split inherent in the structuring and curving of social space. Without making this split of negativity part of the picture, significant shifts in the structure cannot really occur. This is what feminism is about; it is not primarily about neutralizing social differences, but about bringing them to light, and attempting to affect the very structuring of the social space. To do something to/ with this divide, and not simply to try to climb to the right side of it (...) *Thinking in terms of the split* can be a good way of orienting us within concrete struggles to determine which ones are worth fighting for" (Zupančič 2019).

2 At least in American English, *trans-* as a prefix is increasingly suggestive of gender transition. In the Ettingerian understanding, which we are focusing on here, *trans-* hints at ethical and feminine-maternal models of building identity beyond body modification such as fragility, permeability, openness to the other, and the ability to move toward the other: "I also wish to emphasise that the matrixial model doesn't promote natalism and is relevant not to mothers only but indicates a substratum in the subject *of any sex and gender*; matrixial affect, sensing, effects, and awareness arise throughout life both during relationships and in solitude" (Ettinger 2019, 186).

3 Here, a correlation of Maturana's and Varela's autopoiesis with Heideggerian concept of "care" [Sorge] (Heidegger 1996, 114) unfolds. Heidegger's Dasein, in its anxious being-toward-death, is primarily characterized by the agency of care, which, like anxiety, orients Dasein to the context and place of the earth – to the condition that is *always already given* and in which the Dasein of the individual human is situated. Heidegger developed the concept of "care" from the lineage World – Dasein – anxiety – death: "the being Da-sein means being-ahead-of-oneself-already-in (the world) as being-together-with (innerworldly beings encountered). This being fills in the significance of the term *care*, which is used in a purely ontological and existential way" (179–180). With "care (...) we used to designate the being of Da-sein in general" (114), and this care is "concerned about that being" (96). The Heideggerian care is thus fundamentally the care for one's own Dasein. In its care, Dasein is a "concerned being-with with the others" (274) with the aim of preserving *one's own existence*.

4 "Bring forth a world is the burning issue of knowledge (...) this *bringing forth* of a world manifests itself in all our actions and all our being" (Maturana and Varela 1992, 27).

5 "Technē, as knowledge experienced in the Greek manner, is a *bringing forth* of beings in that it brings forth what is present as such out of concealment and specifically into the unconcealment of its appearance; technē never signifies the action of making" (Heidegger 1993a, 184).

6 Translated as: "polyvocity (...), collective arrangements (...) [and] open systems."

6 Aerials of Subreality

Given that the fragile, trans-connected I breathes for the other, consenting to be permeated and concerned by the other, it eludes static comprehension. Instead, it unveils itself in motion, floating, merging into one another – akin to dust particles wafting through the air, intermingling anew at every moment. The carrying paradigm thus transcends the mere subjectivity of the individual. It unveils a mutually carrying *aerial domain*, a realm beyond enclosure (atmosphere) and flatness (cover), transcending predominant spatiotemporal constructs. It is a space-time that can be characterized as *artistic and poetic*, crafted by floating subjects breathing for and with each other.

*

Martin Heidegger's notion of the *fourfold* [Geviert], as presented in his 1949 *Bremen lectures*, can serve as a starting point for our considerations of such an artistic airspace of carrying. Heidegger initially saw the fourfold as a fundamental expression of the world itself: the fourfold involves interweaving the four instances of "earth and sky, divinities and mortals" (Heidegger 2012, 18). These four agencies are "simply entrusted to each other" so that nothing less occurs in their interplay than "the worlding of world" (18). Heidegger stated that the instances of the fourfold are not yet complete in any way. He illustrated this through the example of the mortal, who has fundamentally oriented his existence towards death by "being-toward-death" (Heidegger 1996, 233) but who, nevertheless, has never yet reached that point where he is willing to die, where he is "capable of death as death" (Heidegger 2012, 17). Heidegger concluded from this: "Hundreds of thousands die in masses. Do they die? They perish. They are put down. Do they die? They become pieces of inventory of a standing reserve for the fabrication of corpses. Do they die? They are unobtrusively liquidated in annihilation camps" (Heidegger 2012, 17). Heidegger noted that people die and are put down in masses. This dying of humans is not the dying of Heidegger's *mortals*, but merely a "horrific and dying death" (17). Heidegger stated: "The human is not yet the mortal" (17). Without directly answering the question of when one becomes such a mortal, the thought arises from Heidegger's statement that not only is man incomplete, but so are earth, sky, and the divine. An Ettingerian understanding

DOI: 10.4324/9781003603672-9

of the world and the cosmos relates to such a not-yet-completed concept of the fourfold, particularly as Heidegger's specific tautology resonates with Ettinger's demand of "humanising the human" (Ettinger 2019, 183). At first, it seems obvious that the Heideggerian interplay of the four instances, "earth and sky, divinities and mortals" (17), parallels Ettingerian thinking. For, just as Ettinger carries within the matrixial theory the claim to link subjectivity and the world in attempting "to shift the field of debate beyond the phallus and to make time to consider the possibilities and implications of the matrixial borderspace" (Pollock 2006, 29), so the Heideggerian fourfold follows an initially similar direction. The possibilities of the matrixial stratum, of jointness-in-difference, lie first in the involvement of a continuously *emerging co-existence* that attempts to weave the phallic into the matrixial state of carriance and jointness-in-difference. In this way, Ettinger pursues conceiving subjectivity as an extended, fragile, and thoroughly relational multiplicity based on the primary communication in the maternal womb space. She attempts to formulate a hopeful and liberating constitution of both subjectivity and the world by pointing to manifold forms of responsibility and relatedness. With the matrixial theory, Ettinger made it possible to create an alternative to androcenic mechanisms, which leads to different ethical and political practices than the barbarities of the anesthetizing present. Finally, Ettinger counteracted the prevailing, thoroughly oppressive, and androcentric psychoanalytic and philosophical narratives by introducing into contemporary discourses a theory situated with/in the world and an affirmative, affective, and involved awareness for a thoroughly transitive and co-poietic view on subjectivity and the planet. With the fourfold, Heidegger followed a similar approach towards the world. To clarify his trajectory, Heidegger focused on the concept of the *thing* that marks an object beyond the modernist Cartesian dichotomy of subject/object, as Heideggerian scholar Andrew J. Mitchell (2015) highlights: "Unlike the self-enclosed object of modern metaphysics, the thing is utterly worldly, its essence lying in the *relations* it maintains throughout the world around it, the world to which it is inextricably bound" (3). In this sense, the fourfold suggests a different view of the isolated constitution of the world, which is concretized in the current androcenic epoch. The world can only make itself immediately clear if human beings as mortals have truly learned to die and no longer see themselves as "pieces of inventory" (Heidegger 2012, 53) of a materialistic machine or regard others as such an object material. Heidegger's fourfold is to be read in parallel with his critique of technology and indeed forms a review of the androcenic planetary construction with its dominant mechanistic forms of thought and action. Instead of emphasizing the autonomy and self-determination of subject and object – human and cosmos – the fourfold points to an *essentially relational worldview* that arises only from the interaction and "gathering" (Heidegger 2001b, 171) of "earth and sky, divinities, and mortals" (Heidegger 2012, 18). And yet, the four instances of the fourfold are not yet complete, "the human is not yet the mortal" (54). In Ettinger's view, the statement would be: the human is not yet humanized. Here, Heidegger and Ettinger meet: the fourfold becomes a threshold, a border area where earthly and cosmic actors and agencies are closely interwoven and transgressing.

*

Both Heidegger's fourfold and Ettinger's threshold ground on an immediate artistic or poetic state. For Heidegger, the fourfold is a poetic condition throughout. Heidegger states that mortals can then truly die, in that they are "in the fourfold by dwelling" (Heidegger 2001a, 148). And dwelling, for Heidegger, is connected with the assurance that "man dwells poetically" (Heidegger 2001c, 212). In precisely this concept of poetics and the deeper etymological root of ποίησις [poiesis], it can now be explained in more detail what understanding Heidegger ascribes to art and what artistic understanding, in turn, is rooted in the Ettingerian matrixial theory. This synopsis further allows us to make visible a difference in Heidegger's and Ettinger's thinking and, with this difference, to reach a point where a novel form of planetary ethics can be anchored. Such ethics, then, is not based on mortality and being-towards-death but proposes *a different worlding* – one that can truly realize "humanising the human" (Ettinger 2019, 183) as a societal form of action. For Heidegger, art is a constant interplay between "clearing and concealing. It is (...) the primal strife in which that open center is won" (Heidegger 1993a, 180). This interplay is not mutually exclusive but creates something *new*, an open center in the tension between clearing and concealing. Furthermore, art is "the embodiment of places (...) Places, in preserving and opening a region, hold *something free gathered around them* which grants the tarrying of things under consideration and a dwelling for man in the midst of things" (Heidegger 1973, 7). According to Heidegger, art thus represents the production of a *place* that is not a determined object but an open condition, an empty space of unlimited indeterminacy. For Heidegger, however, such a state of art is never related to an emptiness in the sense of Cioran's void but always creates something *new*: "emptiness is (...) therefore no failure, but a bringing-forth (...) Emptiness is not nothing. It is also no deficiency. In sculptural [artistic] embodiment, emptiness plays in the manner of a seeking-projecting instituting of places" (7). While for Ettinger, a sense of bringing forth must always be linked to wit(h)nessing the other, for Heidegger, the emptiness taking place through the dialectic of clearing and concealing, and, on a larger scale, in the four instances' interplay in the fourfold, always means a poiesis of creating the new – "the worlding of world" (Heidegger 2001b, 178). With this Heideggerian view of art, which makes it possible for the mortal to enter the fourfold – that is, the world – the analogy to a creative process is obvious, which Donna Haraway has only recently applied as a groundbreaking concept for her conception of the world: indeed, Heideggerian worlding in the fourfold could also be described in contemporary terms as "collectively-producing systems that do not have self-defined spatial or temporal boundaries. Information and control are distributed among components. The systems are evolutionary and have the potential for surprising change" (Haraway 2016, 33). Haraway called such a procedure "sympoiesis" or "making-with" (5), thereby distinguished it from "autopoietic systems [which] are 'self-producing' autonomous units 'with self-defined spatial or temporal boundaries that tend to be centrally controlled, homeostatic, and predictable'" (33). Although the Heideggerian fourfold shows a close affinity with the relationality of Ettinger and the matrixial claim of "humanising the human" (Ettinger 2019, 183), this affinity begins to dissolve upon closer examination. For Ettinger's matrixial theory "is not about the new" (Ettinger 2018), not about a *new* birth as in "collectively-producing systems (…) [which] are evolutionary and have the potential for surprising change" (Haraway 2016, 33), but grounds in the statement that "we have been already born" (Ettinger 2018). The place where wit(h)nessing blends into witnessing and the transformation takes place "from proto-ethical aesthetic revelation to subjective agency" (Ettinger 2019, 197) is thus a different place than the place of Heideggerian emptiness and Haraway's sympoiesis. Although the worlding that Ettinger proposes is equally connected to art, i.e., poiesis, this is not a place of surprising, phallocentric novelty but rather a fragile, affective, and feminine-maternal fabric. If "humanising the human" (183) aims to succeed, *the Heideggerian fourfold must be fundamentally expanded and mutated in the direction of a co-poietic and matrixial inter-/intra-action*, an artistic practice that enables a caring adoption of the traumas and atrocities of the androcenic present. It is not about leading mortals to true death but working "with the moment just before death" (Ettinger 2018), which may suggest a fundamentally different ethical path.

*

For Ettinger, artistic practice does not lie in sympoiesis, as in Heidegger's fourfold and his concept of worlding, but in the concept of a consistently contextual and transitive copoiesis. Here, it is not novelty, innovation, or evolution, but the notion of *trauma* that is of central importance: "I am going to propose a passage in art today from Reality, Concept, or Phantasm – to Trauma, and hence a shift for thinking art in matrixial terms" (Ettinger 1999a, 15). The passage that Ettinger proposes here is not the passage of the Heideggerian fourfold, but the passage in which "an encounter with trauma that is shared in trans-subjectivity between the artist and the world" (15) takes place. Here, an artistic practice emerges that mutates the Freudian concept of cultural action as "civilized activity" (Freud 1959, 187) and embeds art's exclusivity, dominance, and aggression in a fragile, interdependent, and responsible fabric. Civilized activity, in Ettinger's view, consists first of "[absorbing] the effects of depth-working of subreal strings and threads" (Ettinger 2015b, 353). This "string-working" (353) means establishing threads and connections with the actors and agencies of the present and the past. In this process, affirmative and aesthetic awareness is formed that is always approaching and emanating from the other(s) and her/his/their trauma. With this approach, Ettinger makes it possible not to anchor ethical action as a self-referential element in a synchronized system of an auto- or sympoietic mechanism but to see it situated in an open fabric that makes tangible what usually remains inaccessible in hegemonic notions of human subjectivity. In such an artistic practice, an encounter is made possible with "a subject-matter even without having witnessed in reality the event that it indicates" (Kaiser and Thiele 2018, 119). A possibility is opened to encounter the other(s) and her/his/their trauma, even if I did not directly experience such a traumatic event by myself. Through the embedding in the passage – this "almost-impossible rapport (…) between me, the Other and others" (Ettinger 1999a, 16) and the artistic "string-working" (Ettinger 2015b, 353) – the transitive constitution of subjectivity becomes manifest: indeed, the human psyche has more connections and linkages to spaces and times than is generally thought possible. When Heidegger spoke of the importance of the poetic and the artistic, this is equally true for Ettinger's "string-working." However, the difference becomes apparent when Ettinger's poetics consists precisely of *not producing a new work*, nor a new form of relational aesthetics, but rather a transsubjective and transobjective resonance that is based not on physical but on affective, trusting, and carrying relations (Cf. Ettinger 2018). Ettinger's poetics is about "the beyond of the 'being-with-oneself'" and about the "being-with-oneself-with-in and in-resonance-with another" (Ettinger 2015b, 365). This is a fundamentally different artistic understanding than that of Heidegger, but also that of Freud. Both Freud and Heidegger were concerned with the creation of new worlds and spaces, with "clearing-away," "clear cut," and made "free from wilderness" (Heidegger 1973, 5) in various forms. On the other hand, Ettinger is interested in existing spaces and contexts, in their stories, characteristics, and traumas. Her approach is not an operation of a phallocentric creation whose sprawl continuously contributes to the androcenic reality of life. Rather, it reveals a subtle and fragile artistic practice whose threads can reach undefined depths: a "work in depth, beyond the abstract pure surface of transcendence" (Ettinger 2018). This transitive understanding of art, which moves from poetics to ethics, also reveals the important aspect of mutual responsibility continuously emerging in such a "depth-working of subreal strings" (Ettinger 2015b, 353). In the transitive fabric, an "affective, emotional cathexis in singular cared-for/caring entities" (Ettinger 2019, 194) occurs, i.e., a mutual engagement combined with an assumption of responsibility and care for the other(s). In this fabric, it becomes clear that "once you carry, you are carried too" (Ettinger 2015b, 355). Subjectivity is continuously determined by the other's otherness, and a constant – often ignored or hardly perceivable – inscription between the other(s) and the self occurs. In Ettinger's understanding, matrixial artistic practice can make these inscription processes visible and tangible. The transitive and feminine constitution of subjectivity now shows that the individual trauma is always a trauma of togetherness or a "transitive trauma" (Ettinger 1999a, 16), which is constantly created by others but – at the same time – is also carried by them. With this transitive and trauma-centered conception of poetics and art, Ettinger advanced to a fundamental re-conception of the beautiful, in which the aesthetic and the ethical, the local and the global, or the individual and the political start to meet.

*

For Ettinger, the beautiful no longer represents a sublime solidity charged with the Freudian uncanny. Instead, beauty is "the access via metramorphosis to the wound of the Other" (21). Metramorphosis – the feminine transformation of one's self through mutual carriance with/for the other – makes the fragile string configurations accessible. In the artistic process, the beautiful begins to emerge when a transsubjective and transobjective resonance is established between the self and the other, and the consciousness is created of being part of the other as the other is part of one's self; in such moments, the painter assumes that "the tableau vibrates" (21), but such a tableau can take on a different form for each human-artist. In any case, there is an extraordinary potential for transformation in such a process, making the trauma of the other visible, accessible, and comprehensible: "traces of a buried-alive trauma of the world are re-born from amnesia into co-emerging memory" (22). In the last instance, however, it is not the artistic work or the process of making art – the tableau, the drawing, or the performance – that is essential. What is decisive is that with this specific process, a passage opens up, allowing it to create a caring, carrying, and hopeful co-existence and co-emergence between actors, spaces, and times previously situated far from one's consciousness. The process and the work can be considered artistic, aesthetic, and beautiful only if they make it possible "to abandon defenses and become fragmented and fragile, to share, absorb and further redistribute fragments of trauma" (22). This is not about a distant, objective sublimity or an indefinite uncanniness but about an affective entry into a dense fabric that connects me with the other(s) and the other(s) with me. In this aesthetic view of coinhabit(u)ation, the beautiful is never a Heideggerian "thing" but a field of possibility and entry point that works through the individual trauma and allows it to branch out into new paths of hope. As Ettinger noted, "it is in creating objects, procedures and events of and for a sick world that the artist becomes both the (...) insane and the doctor" (16). However, the artistic process is not only about the individual psychic sphere. The transitive fabric is not only human but also more-than-human. A multidimensional tissue is stretched out here, in which "traces of the trauma of the world and of others" (16) are processed, inscribed, and made tangible: a *psycho-planetary continuum*, where artistic string-working contributes to the construction of a fragile carrying architecture. In this sense, it becomes clear that artistic thinking and doing "has a vital role to play in curing our environmental ills" (16).

For Ettinger, art becomes a "transport-station of trauma" (Ettinger 2016a, 151) when it understands itself as a passage in which the strangeness of the other(s) is thought together with a conscious decision for and an affirmative entry into this otherness itself. To be without and within. If the gift of this passage is accepted, the aesthetic realm then transforms itself into an ethical and political dimension. It does not remain exclusive and excluding ("the beautiful," "the sublime") but rather unfolds as an integrating and transitive practice of exchange and of taking mutual responsibility. With this artistic and aesthetic horizon proposed by Ettinger, it is possible to form an Ettingerian respiratory practice that turns to the eco-political contemporary and makes its trauma visible and sharable. In this matrixial form of respiration, the human is on the way to becoming humanized: "Not only am I concerned by my own wound (...) but I am also concerned by the wound of the Other" (Ettinger 1999a, 18) means becoming-interwoven into the traumatic layers of the planet, whose total otherness gives way and manifests itself as "my non-I(s) which are transmitted to me" (18). This psycho-planetary relationality implies a fundamentally different conception of breathing patterns, which puts both the hegemonic biological and philosophical narratives on hold in favor of a *respiratory practice that is connected to and visible through artistic practice and co-poietic "string-working"* (Ettinger 2015b, 353). In Edvard Munch's (1863–1944) woodcut *Towards the Forest I* (Figure 6.1),

Figure 6.1 Edvard Munch, *Towards the Forest I*, 1897, Color Woodcut. 60.1 × 80.1 cm (23.6 in × 31.5 in). Munch Museum Oslo.

it is precisely this string-like aesthetic that leads Munch's work of art away from a mere object and makes it a key to an ethical and political practice of "enjoying encounters of I(s) with non-I(s)" (Ettinger 1999a, 18). Here a new kind of interweaving between the self and the other manifests itself, which, beyond representation, points to a *matrixial knowledge*, the development of which seems to be indispensable for dealing with the androcenic barbarities of the present anesthesia. In this performative dialogue between the human and the more-than-human, as suggested within Munch's woodcut, Ettinger's matrixial theory is thus immediately "channeled toward different fields," and a procedure is set in motion that "labor[s] toward humanising the human" (Ettinger 2019, 183). A passage opens up that offers a hopeful alternative to the collective and abysmal striving towards the "horizon of being-toward-death in anxiety" (212).

The horizon delineated here may be construed as *aerials of subreality*, within which the meteorologies of the matrixial are unveiled: "Something of it will arrive, to me and to-from the cosmos. You are, I am, grains of dust in the eyes of the vast glowing vault. I long to be carried but I must carry you" (Ettinger 2015b, 354). What becomes discernible in Munch's woodcut is no longer the dichotomy of subject and object nor the establishment of a clearing – an "open center" – where the world and life commence. A different dimension unfolds that transcends "clearing and concealing," surpassing enclosure (atmosphere) and flatness (cover): a reciprocal, non-violent permeation, a becoming affected by the other, borne from the awareness that the other inherently embodies a part of my self. In his composition, Munch portrays a loose, non-mechanized intertwining that is divergent from the predetermined processes of auto- and sympoiesis. Within this string-working of I and (non-)I, a breathing space emerges that not only inhabits the present but also extends into the archaic and the future: the space of the *subreal*, which Ettinger introduces with the definition: "The significant exchanges occurring in the duration of real and symbolic carriages evolve at the level I term the subreal. Transjectivity truly transpires, in subreality. We discern resonance beyond the senses" (Kaiser and Thiele 2018, 110). Such resonance is precisely evident in the interplay between humans and nature in Munch's woodcut, for which Ettinger's reflection of the artistic product also applies:

> Grains of matter, ashes of light and shadows of void relate to one another and are injected backward or onward to the cosmos. With-in but with-out of such beauty, copoietic consciousness emerges; we are accessing the subreal copoietic string knowledge of growth and decay, utopia and apocalypse, evolution and entropy with-in cosmic time-space in its move toward the sub-symbolic web of threads.
>
> (Ettinger 2015a, 350)

The emerging breathing dimension is not biological, mechanical, symbolic, or abstract. It does not concern the creation of a new work in an "open center" but instead entails a genuine *aesthetic* dimension insofar as aesthetics, in its understanding as "pertaining to perception," appeals to the sensibility for the other(s). Such a perspective is reflected in a subreal characterization as "a net of strings of aesthetic and human connections (...) It fosters connections of 'co-emergence': 'I feel in you,' 'you think in me,' 'I know in you,' and so on, whereby subjective existence finds articulation through one another. Only art can achieve such an encounter" (Ettinger 2016c). The breathing space thus created is perceptual rather than spatial, occurring not cyclically but aesthetically; the breath flows into each other, is perceived by the other, absorbed, taken over, and carried forth so that the fundamental practice in such a spatiality may be described as "breath-floating" – a floating of living particles, grains of color that blend, overlap, and actualize the virtual. Such a practice varies for each human and more-than-human grain and is differentiated in its dynamics. In breath-floating, manifold expansions, rhythms, positions, and velocities are unveiled: exhalation drifts toward the other and is carried by the other; inhalation brings the other to me and is carried by me. Air drifts of the other permeate me, comprising not only present air drifts but also traces of a traumatic past or a bleak future. Through breathing as a color particle, I inscribe myself into other particles, engendering a different chromaticity and novel forms of symbolization:

> Subreal elements and strings resonate with-in and in-between human beings humanely in touch with the cosmos. Invisible threads are coweaved, they find venues into the visible plane and into consciousness. When this resonance is sensed or is rather trans-sensed and stirs our passions and reaches to our desire to dwell with this mystery long enough, the imaginary thus opened to and by the subreal enlarges the symbolic.
>
> (Ettinger 2019, 194)

*

Theorizing the aesthetic breathing space necessitates concrete reference to Ettinger's painterly work centered around people, primarily women and children, subjected to humiliation and violence during wartime:

> Painting works slowly. It allows us to enter the space of the trauma of others and of our own with neither fight nor flight, and to dwell in its resonance. Painting is about bringing into visibility that which is not ordinarily visible, including the forms that violence takes. Painting produces a suspension in time. It not only makes us confront the atrocities of the past, but provokes how we see and feel about the present moment.
>
> (Ettinger 2016c)

In the aesthetic airspace, manifested on a smaller scale in every "artworking" (Ettinger 1999a, 19), and more specifically in every painting, the "impossibility of not-sharing with the other, nature, and the cosmos" (Ettinger 2019, 200) is realized; such participation is only possible within such *aesthetic* airspace and via an *aesthetic* breathing practice, actualizing a different, non-phallocentric symbolization beyond the realm of bliss: "The painting itself as a new image-body-matter activates a space in itself and in me. A subject appears" (Ettinger 2015b, 355). For Ettinger, the process of "artworking" is a practice of communion "with the other, nature, and the cosmos" where

> different traces and vibrations from different encounters at different moments slowly accumulate over the years on the same canvas. But a trace can also be dispersed among different canvases. I am always working on several paintings at the same time. I am scanning the canvas, tracing, erasing, imprinting, dispersing photocopic dust, smearing and spreading pigments and ashes, snaring a moment, snatching a trace, slowing an affect, transferring a line, connecting a cord, and sifting some grains. I secure the exchange, transmit what emerges from it, let it slip, deliver the lines and phrases, strip them naked, give them a new measure of freedom, disturb the cords, let them appeal, engrave their protest, lay the trace bare again, shutter it, strengthen it, destroy it, efface it, effuse its remainders, charge the traces with new intensities coming from myself and from the other, from outside and from within, and I listen carefully. Again and again, I let the remnant of the process pass, transformed, to the other; I let it be entangled in a web of chords; I draw out a cord, stretch it back, give it a visual voice, protect that voice.
> (Ettinger 2001a, 44)

This aesthetic process is characterized by a continuous movement forward and backward, as well as by an affected perception aiming to document, render visible, and retain the traces that blend into one another. This is not merely a practice for its own sake but primarily *for the other within me*, a breathing process that secures my own life and, foremost, the life of the other. I do not breathe for myself; I breathe for and with the other, who manifests herself/himself/itself as the (non-)I: no longer alien but inherently belonging to me without, however, being appropriated by me, thereby transcending any phallocentric viewpoint:

> Therefore, what we discover are new links, operating outside as if they come from the inside, operating inside as if they come from the outside and linking inside and outside. That's why we can't discuss total separateness, symbiosis or non-differentiation. Total separateness, a "normal" phenomenon in the phallic arena, is pathological in the matrixial

> sphere. Transgression of borders, a perversion for the phallic arena, is not perverse in the matrixial sphere. There is "separation-in-proximity." (50)

The other is no longer an exclusive threat: "The other is a partial-subject in a shareable assembling. It has its threatening aspects, it has its attuned aspects, it attracts and it rejects, it opens distance in proximity, it cannot be entirely known" (55). However, every breath is also tied to a decision to assume such responsibility, to interrupt hegemonic autopoietic, symbiotic or biologistic breathing patterns, and to embark on this passage towards the subreal aerial dimension of breathing for the other as a partial subject: as such, it always entails "a choice, an option, an offer, a gift" (Ettinger 2011, 7). In its trans-subjective and transobjective intertwinement, the subject retains the freedom to withdraw from or accept this offer. The aesthetic sensibility for the other thus merely provides the *possibility* for a space, delineating a path upon which the subject must always decide anew. This engenders a tension oscillating between engagement and withdrawal. Yet, in every moment wherein the decision is made to diverge from the mechanics of biological respiration, the "subject is now transformed, informed by matrixiality, assuming carriance" (Ettinger 2015a, 343). This transformation occurs upon the decision to "turn once more from the level of the symbolic to the level of a corporeal gestation as it is experienced by a female gravida" (Ettinger 2019, 197). The choice for or against the suggested passage perpetually embodies a choice for or against the feminine matter of concern of the subreal psycho-planetary intertwinement with the other; it represents the decision to *breathe aesthetically* – i.e., with and for the perceived other. Only through breathing aesthetically can the subject emerge as a pivotal actor, a carrier, initiating or advancing "labor toward humanising the human" (183). Only then, when in breathing "different traces and vibrations from different encounters at different moments slowly accumulate over the years," the human is truly the mortal in the Heideggerian sense. While it is imperative to cultivate an interplay between "earth and sky, divinities and mortals," an extension of Heidegger's fourfold becomes necessary: For, such an interplay must never dialectically unfold the open center, the "clearing" (Heidegger 1993a, 180), i.e., a "total separateness," but rather endeavor to absorb traces and resonances of the other, layering them into a closely knit, delicate "web of chords": "the first moment of what is going to be the work of art is already an accumulation of traces and dust, a passage from the Other, dead or alive, inside myself and out there in the world" (Ettinger 2001a, 51). It is vital to acknowledge that such "aerials of subreality," unlike hegemonic atmospheres or surfaces, remain undefined, only formed anew with each breath – with each perception, absorption, and affection – and are never repeated – as with the artworking process of painting. Nonetheless, aesthetic breathing does not unfold contextless: "There is never a moment when painting begins with nothing from a blank page" (51). It revolves around accumulations rather than the preservation of established static systems: about breathing that engenders a healing dialogue wherein the "echoes of different conversations reverberate. We can talk about hollow spaces of affection" (58). These "hollow spaces of affection" represent a distinct dimension from Heidegger's "clearing." They constitute an airspace of "co-response-ability" (54), of being affected and *floating-into-each-other*, facilitated by rûah's eternal breath. Breathing, as a fundamental aspect of human existence, begins to entail other prerequisites that may further lead from aesthetic breathing to ethical action and a different contemplation of the relation between the human and the more-than-human realm.

Part III

Metreorologic Speculations – Toward Ettingerian Environmental Ethics

7 From Respiration to Trans-Spiration

In elucidating the procedures of floating-into-each-other within the "aerials of subreality," it becomes imperative to highlight the respiratory dynamics inherent in Ettinger's concept of "co-response-ability" in regard to their societal and political implications. What does this aesthetic respiratory form, described by Ettinger as "string-working" (Ettinger 2015b, 353) or "art-working" (Ettinger 2019, 191), embody in its actual enactment, always entwined with a specific socio-political milieu? What consequences may this alternate mode of respiration, emerging in the affective and co-poietic encounter between the I and its "non-I(s)" (Ettinger 1999a, 18), entail? What political impulse might arise from such a breathing rhythm that resists totalization and phallocentric dominance in a fragile yet profoundly significant manner? I suggest that the rhythm of the breath drifts, floating into and carrying one another, manifesting beyond predefined sensory perception, chronos, and spatial definitions within a subreal fabric of strings, address and unfold a specific *virtuality* intrinsic to every human being but often effaced within phallocentric territories of control and anesthesia. Accessing this virtuality is not straightforward; it lies concealed, necessitating initial acknowledgment, as aptly characterized by Gilles Deleuze:

> In short, the virtual is never subject to the global character which affects real objects. It is – not only by its origin but by its own nature – a fragment, a shred or a remainder. It lacks its own identity (…) Conversely, these virtual objects are *incorporated* in the real objects. In this sense they can correspond to parts of the subject's body, to another person, or even to very special objects such as toys or fetishes. This incorporation is in no way an identification, or even an introjection, since it exceeds the limits of the subject.
>
> (Deleuze 1994, 100–101)

Virtualities augment and potentially transcend individuals, albeit always remaining mere traces. In actualizing these virtualities and making these invisible traces perceivable *through the practice of floating-into-each-other*, subjectivity – and humanity – unfolds: "Breathing, not only air, is what bestows subjectivity on man and provides the grounding for his psychological

DOI: 10.4324/9781003603672-11

as well as her spiritual dimensions" (Škof 2015, 134). This statement marks a pivotal argument against a biologistic view of respiration and an entirely symbiotic world comprehension, and is reminiscent of Emmanuel Levinas' thought. In a poetic form, Levinas speculates on subjectivity, symbolizing dimensions akin to Ettinger's "aerials of subreality":

> That the emptiness of space would be filled with invisible air, hidden from perception, save in the caress of the wind or the threat of storms, non-perceived but penetrating me even in the retreats of my inwardness, that this invisibility or this emptiness would be breathable or horrible, that this invisibility is non-indifferent and obsesses me before all thematization, that the simple ambiance is imposed as an atmosphere to which the subject gives himself and exposes himself in his lungs, without intentions and aims, that the subject *could be a lung at the bottom of its substance* – all this signifies a subjectivity that suffers and offers itself before taking a foothold in being. It is a *passivity, wholly a supporting.*
>
> (Levinas 1991, 180)

Levinas suggests that in the emptiness of space, where only the traces of a traumatic past remain, a wind is blowing, a virtual and incessant floating-into-each-other, only barely apparent, suppressed in its weakness by the androcenic regimes of enclosure and flatness. Levinas alludes to a biblical interpretation of this faint voice, scarcely allowed to actualize itself in the human: "The voice of him that crieth in the wilderness, Prepare ye the way of the Lord, make straight in the desert a highway for our God" (Is 40:3). The prophet later laments the chosen people's disregard for this gentle voice: "when I called, ye did not answer; when I spake, ye did not hear; but did evil before mine eyes, and did choose that wherein I delighted not" (Is 65:12). It appears to be a fundamental narrative of human nature to disregard the subreal airspace and the respiratory virtuality of floating-into-each-other, despite their concealed presence. According to Levinas, however, humans are fundamentally constituted by their lungs, their capacity to embrace the other's breath within their innermost being: "the caress of the wind (...) penetrating me even in the retreats of my inwardness" (Levinas 1991, 180). Such openness to the other defines a subjectivity that "suffers and offers itself."

*

The biologistic perspective on breathing posits that "organisms are less self-enclosed, autonomous individuals than communities of bodies exchanging matter, energy, and information with others. Each breath connects us to the rest of the biosphere, which also 'breathes,' albeit at a slower pace" (Margulis and Sagan 2000, 23). In this symbiotic comprehension of breathing, better termed *re-spiration* – derived from the Latin re-, meaning "again," and spīrō, signifying "to breathe, blow" – there always exists a perpetual cycle and a continual renewal. This re-spiration entails a constant repetition, replicating previous bodily movements, where each breath echoes the preceding one. It engenders a perception of continuity as if each breath were a replica of the last. To elucidate the subtleties of floating-into-each-other, it proves beneficial to think about the notion of *rhythm*, alongside the concepts of *territorialization* and *deterritorialization* prominently introduced by Deleuze and Guattari. These conceptual tools shed light on the inherent virtuality of floating-into-each-other as accessible to every human through her/his lungs – offering a pathway toward an alternate breathing rhythm beyond the confines of repetitive re-spiration. In general, rhythm must be distinguished from the mere meter. Whereas meter entails mere reproduction, rhythm remains unbounded, subject to transformation and variation, yielding a continuum of silence, pause, and sound in equivalence and simultaneity. Prevailing hierarchies or patterns are initially foreign to rhythm. Such a nuanced understanding of rhythm contrasts with the hegemonic view of time as *chronos* and is closer to a continuous flow of *aion*. Aion resembles the "simple ambiance (...) imposed as an atmosphere to which the subject gives himself and exposes himself in his lungs" (Levinas 1991, 180), embodying a flowing continuum where intensities form and transcend predetermined patterns. In opening myself to the other, my corporeality transforms; subjectivity – and thus, humanity – emerges within an unpredictable context, distinct from the determinism of biological symbiosis. For Deleuze and Guattari, pivotal moments of crystallization – indeed, the genesis of humanity – are intertwined with the emergence of an uncharted, yet-to-be-explored territory. These territories arise from rhythms, themselves born from the depths of aion – the ineffable "emptiness of space (...) filled with invisible air" (Levinas 1991, 180). Aion, far from negating all order, constitutes a primordial medium akin to the biblical "wilderness" and "desert" (Is 40:3), from which emanates the whisper of God, the omnipresent rûah. From this aion, indeterminate milieus and rhythms take form, initially characterized by periodic repetition. However, an aperiodic rhythm prevails between two or more milieus or within sub-milieus of the same milieu. Such rhythm is exemplarily resembled by the pulsation of the heart, where the periodic contraction of its muscles produces repetition, yet the rhythmic continuum among other bodily parts or between the heart and the world undergoes constant flux:

> A milieu does in fact exist by virtue of a periodic repetition, but one whose only effect is to produce a difference by which the milieu passes into another milieu. It is the difference that is rhythmic, not the repetition, which nevertheless produces it: productive repetition has nothing to do with reproductive meter.
>
> (Deleuze and Guattari 1987, 314)

As elements within the milieu acquire qualities shaped by the rhythm that animates them, milieu and rhythm metamorphose into a territory, marking *territorialization*. Conversely, de-territorialization occurs when an element transgresses its original milieu function to assume a new territorial role, no longer tethered to its prior function. This process mirrors the behavior of the *tooth-billed bowerbird* during mating season, as it marks its territory by collecting leaves and arranging them on the ground – undersides facing upwards – inviting others into this space (Cf. Bogue 2003, 19ff.). Once having been an element of an original milieu, each leaf becomes a transgressive rhythmic component in de-territorialization through the bird's behavior. The leaves cease to be merely leaves, instead pointing toward *something else* – a new space, an undiscovered dimension, an anticipation of the other's arrival.

*

How might this theory of territorialization and deterritorialization be transposed onto the dynamics of breathing and the contemplation of a breathing practice transcending periodicity and repetition? Initially, every human may be seen as a defined "existential territory" (Guattari 1995, 17) emerging from the aion of the wilderness. Each moment of childbirth, each inaugural breath, can be construed as an act of territorialization, shaping bodily territoriality through initially repetitive, biologically determined rhythms and milieus. Yet, what proves pivotal in this territorialization is the acknowledgment that every human – indeed, every living being – arises from the wilderness of aion, *from the flow of the rûah* and thus bears such history as virtual traces within: "And the LORD God formed man of the dust of the ground, and breathed into his nostrils the breath of life; and man became a living soul" (Gen 2:7). Sustaining life necessitates the continuation and repetition of breathing after birth, ensuring the persistence of individual physical rhythms that constitute an organismically functioning corporeality. However, this bodily vitality represents merely one facet of aliveness. Relying on biology and existing respiration patterns alone cannot sustain life. Referring to Deleuze and Guattari, adherence to solely biological respiration risks reducing the human – a territory revolving in on itself – to a state of impoverishment and eventual collapse. Such impoverishment resonates with the (post-)anthropocenic present, characterized by prevailing internalization and flatness, fostering the illusion of eternal respiratory sufficiency. Deleuze and Guattari underscore the ethical significance of de-territorialization, wherein humans become aware of their lung function, inviting the other – the stranger – into their corporeal being, into the enclosed "existential territory" as Levinas expresses it, transforming into a "subjectivity that suffers and offers itself before taking a foothold in being" (Levinas 1991, 180). A deterritorializing subjectivity "that suffers and offers itself" signifies transcending respiratory practice and actualizing its potential of floating-into-each-other. Deleuze and Guattari claim that

> deterritorialization is always double, because it implies the coexistence of a major variable and a minor variable in simultaneous becoming (the two terms of a becoming do not exchange places, there is no identification between them, they are instead drawn into an asymmetrical block in which both change to the same extent, and which constitutes their zone of proximity).
>
> (Deleuze and Guattari 1987, 306)

De-territorialization never occurs in isolation; it is invariably entwined with the other. When breathing transcends mere biological repetition and becomes *inviting* – as with the enticement of the bird's leaves – breathing deterritorializes itself, no longer confined to itself but manifesting as the coexistence of two singularities – two breathing entities encountering – the "coexistence of a major variable and a minor variable." Notably, this reciprocal breathing does not entail appropriation of the other; the de-territorializing procedure of breathing always remains asymmetrical, approaching and withdrawing from each other.

*

The human being embodies the potential for lifting repetitive respiration. Yet, unfolding this potential is contingent upon dismantling corporeality defined solely in biological terms – a dismantling far from harmful or even destructive. However, "dismantling the organism has never meant killing yourself, but rather opening the body to connections that presuppose an entire assemblage, circuits, conjunctions, levels and thresholds, passages and distributions of intensity, and territories and deterritorializations" (Deleuze and Guattari 1987, 160). Indeed, the de-territorialization of breathing is twofold: a journey into one's past, relinking subjectivity with traces of the divine rûah that breathed humanity into existence. Simultaneously, it points toward a future where de-territorialization envisages the condition for the continuation of aliveness. *Without de-territorialization of breathing, without the incorporation of the other's breath into one's lungs, humanity cannot persist*, and if it does, only as "living deads" within the regime of phallocentric dominion. Breathing transcending its functional essence, opening itself to the other, partaking in the other's trauma can be formulated as *trans-spiration* (derived from Latin trans, meaning "across, beyond," and spīrō, meaning "to breathe, blow"); trans-spiration, quite literally, signifies "breathing across," surpassing symbiosis, appropriative relationality, and biological systems. Trans-spirational breathing transcends the confines of one's corporeality: as Deleuze and Guattari aptly put it, it is "always double," entwined in an asymmetrical "simultaneous becoming." It anticipates the other, akin to the *tooth-billed bowerbird* awaiting a potential mate (even without assurance of her arrival). Trans-spiration embodies the dual nature of breathing – biological *and* ethical – understanding that humanity simultaneously embodies individuality (territorialization) and relationality (de-territorialization). Engaging in ethical breathing through trans-spiration involves more than merely recognizing the other's presence, going beyond integrating them into predetermined respirational frameworks. At its core, trans-spiration advances into the essence of breathing, transforming subjectivity into a state of "suffering," "passivity," and "supporting" (Levinas 1991, 180). What precisely does this notion of a *breathing that suffers* entail?

*

Referring to Levinas, trans-spirational breathing inherently embodies *asymmetry*. The other – the "major variable" or "minor variable" – possesses a respiratory system distinct from our own, characterized by different rhythms and milieus. Levinas vividly portrays this alterity, the inherent dissimilarity between one's own and the other's respiratory systems, through the concept of the *face* – an ever-changing visage that gazes upon us, communicates without words, and beckons to be recognized and embraced: "The Other remains infinitely transcendent, infinitely foreign; his face in which his epiphany is produced and which appeals to me breaks with the world that can be common to us" (Levinas 1969, 194). In Levinas' understanding, every thought does not proceed from me, from the sameness of my self, but from the other, who always remains and must remain different and foreign. The other cannot be appropriated; she/he is radically a-synchronous with my self; the other breaks into my world, is never a fellow "Thou" (as with Martin Buber), but always a distant "Vous." If I put the other on my level, I did violence and oppression to her/him/it. Levinas goes so far as to claim that the other cannot be understood either, as understanding presupposes objectification of the other, which renders the encounter no longer ethical. The encounter is ethical only when it departs from the face of the other and its inherent expression, "Do not kill me! – Take care of me!" According to Levinas, it is the ethical obligation of every human not to withdraw from the other, not to abandon her/him/it, but to strengthen her/his/its weakness, to take a carrying responsibility for her/him/it, and to leave aside any destructive power of self(ish)ness. I owe it to the other to hear her/him/it, to carry her/him/it, and to suffer with her/him/it. This claim goes beyond sympathetic empathy. The encounter with the other transforms from a "way of being with the other" to a "way of *being for* the other" (Levinas 1999, 105). The other who chooses me by looking at me, by his face, cries out for my attentiveness, often inaudibly, incidentally, seemingly insignificantly. Eric Severson (2011) has pointed out that such a cry often resembles a *whispered prayer*, a so-called "deësis" (from Greek δέησις, meaning "request," "entreaty," or "prayer"), another form of speech: "even when lips are sealed, the face of the other 'speaks' a language very different from (...) everyday banter (...) The widow, orphan and stranger suffer, and their suffering is a cry from beyond the structures of everyday language" (Severson 2011, 20). Perceiving the other means, on the one hand, to perceive this cry in the other human, who manifests and often remains hidden and unheard by spoken language. On the other hand, it is my obligation to hear the *hope* embedded within the other's spoken words and to "take that place which she/he can no longer take by her/his own strength, where therefore nothing can be made good, and thereby to help her/him" (Böttigheimer 2006, 54, translation mine). In a relationship that is "for the other," that truly carries her/him/it, "the personhood of the other is not necessarily annulled, but on the contrary is only truly released" (57). Carrying the other *vicariously* does not mean robbing the other of self-determination but rather awarding her/him/it freedom: "Unlimited vicariousness does not mean loss of life, but true incarnation" (56). The expression directed at me from the face of the other serves as an important gateway to grasp the significance of trans-spiration. By embracing the virtual, often faint and scarcely perceptible breath of the other, a reciprocal carrying, a turning toward and acceptance of the other unfold – "wholly supporting," wholly carrying. The de-territorialization of re-spiration unveils nothing but the assertion of true humanity: the reorientation of one's territoriality toward the other and the recognition that one's territoriality can only be carried by the other. However, in trans-spiration, the other is not exploited for personal ends; rather, preserving one's vitality occurs only when one accepts the suffering of the other, when the secluded self advocates for the other, essentially doing good unto her/him/it. This good is never self-serving; it materializes only when the supremacy of the self relinquishes, when it endures and shifts from active intentionality to passive extentionality – when one's own prayers fade in importance, eclipsed by the prayers of the other or the prayers for the other. Trans-spiration embodies an inherently restless breathing, persistently disrupted by the presence of the other. The notion that subjectivity is solely forged via a homeostatic equilibrium, as proposed by biologistic perspectives, proves illusory in trans-spirational practices. The longstanding assertion that "the goal of life is global homeostasis, and toward this end it transforms the planet into what might be viewed as a single self-regulating organism" (Roszak 2001, 145) is unmasked as a phallocentric ideology of domination that excludes any alterity, any relation with the approaching other; such truth-claiming theories are instead revealed as cynical, far removed from true vitality. Genuine "incarnation" is only realized beyond rest, beyond a homeostatic ideal: "The identity of the subject is here revealed not through self-assured repose, but through a *restlessness* that propels me beyond the core of my substantiality" (Levinas 1991, 142). Restlessness – interpreted as becoming-unsettled by the other – thus emerges as a main factor of trans-spirational breathing, positioning itself in critical opposition to the historical stance of breathing in Indian philosophy and the ideology of bliss prevalent in the (post-)anthropocenic era. For instance, Patanjali, the writer of the Yoga Sutra (exact life dates unknown, circa between the 2nd century BC and the 4th century AD), posited that the objective of yoga is "the dissolution of the agitations of consciousness," achievable through "the expulsion and retention of breath" (Yoga

Sutra, chapter I, verse 34). Such dissolution of agitation through breathing diametrically opposes a trans-spirational perspective, indeed calling for an ethical analysis, particularly in light of the predominance of yogic practices in contemporary atmospheres of anesthesia. Drawing from Levinas, it can be asserted that vitality and attaining full humanity are inconceivable without restlessness. Humanity is incomplete without the agitation of the other, who approaches and confronts me with their faint drifts and silent outcries. Gaining awareness through a mere dissolution of agitations of consciousness might lead to the realization of barbarism.

*

The drifts of the other do not attack my lungs like a raging fire but unfold in a manner akin to the delicate descent of snowflakes. Paul Celan eloquently captured such a gentle trans-spirational encounter in his poem "Mit wechselndem Schlüssel":

> With a changing key / you unlock the house in which / the snow of secrecy drifts (…) When the key changes, the word changes / which is allowed to drift with the flakes. / Depending on the wind that carries you away / the snow gathers around the word.
>
> (Celan 1964, 36, translation mine)

The respiratory drifts emanating from the other reverberate, each time unlocking one's dwelling of subjectivity – the "existential territory" – in a varied manner. Subjectivity, which absorbs the echoes, blows, and words of the other, allowing itself to meander within their drifts, partakes in the other's trauma as she/he/it unlocks me: "you unlock the house in which / the snow of secrecy drifts." Celan elucidates a pivotal aspect of trans-spiration: the breath-drifts of the other not only blow toward and through me but also possess the capacity to unlock me – provided I transcend the cycle of respiratory repetition and remove the barrier impeding the other's unlocking. Hence, de-territorialization aptly translates into *unlocking*, a readiness to welcome, akin to holding a door ajar, awaiting the other's ability to unlock it with her/his/its "variable key." In trans-spiration, humanity abandons both vertical, phallocentric dominance and horizontal, homeostatic hegemony; the trans-spiring, un-locked human is essentially adrift and floating. No longer confined to an upright posture or trapped within dual modes of movement, she/he traverses realms high or low in a perpetual condition of motion and flux. Dominance dissolves; corporeality is now composed of the absorbed breath-drifts of the other, of traces left by other trans-subjects and trans-objects; the confined "existential territory" dissolves – opening to the other and becoming unlocked by the other.

*

Such trans-spirational practices may also be comprehended through the lens of the feminine, as in the matrixial encounter with the other, "the individual subject is infused with the knowledge of its own transjective and transsubjective links" (Kaiser and Thiele 2018, 110). In trans-spiration, the subject enters the transition "from objectal biological living substance to a becoming-pre-subject as it is intuited by a female subject" (Ettinger 2019, 195). This transition is a thoroughly female one, for in the body-psyche of the female human being there are, always and inherently, traces of subtle gradations between life and non-life, such as "processes of birthing, disbirthing (miscarriage), unbirthing (abortion), a-birthing (childlessness, childfreeness), and birth" (184). Contrary to identarian narratives, the trans-spirational passage of jointness-in-difference follows trajectories of a queer aliveness – where life, non-life, almost-life, and death participate and interweave. Based on the observation that the restlessness initiated by the other, and the openness for the transjective and transsubjective drifts of the other, promotes and ensures liveliness, Ettinger also speaks of a "seduction into life" in the matrixial encounter (Ettinger 2015b, 354), unfolding a matrixial eros "for which the language of germ cells and the idea of the principle of life's longing for its inanimate and even inorganic origin is unfit" (Ettinger 2019, 190). Instead, the matrixial eros follows a process of "caring-carrying-sustaining-bearing-lifting-subjectivizing" (Kaiser and Thiele 2018, 108), which implies a differentiated approach to the question of (sexual) identity. Consciousness around the transsubjective and transobjective links emerges through the becoming aware of the subjectivity-as-care-carrier, as a possibility of openness, affectivity, and receptivity in the human – beyond dualistic standpoints, but essentially floating. At this juncture, it is evident that trans-spiration epitomizes an embodiment of the Ettingerian matrixial theory, serving as the core of a practice that transcends mere physicality and extends further into the political realm – a concept aptly named *Matrixial Breath*.

*

Ettinger posits that the expressions of the caring-carrying libido do not only occur between *human* I and non-I(s) but also with the non-I(s) of planetary matter, physical forces, and atmospheric intensities – ultimately joining the contingencies of the cosmos, in all their ambivalence, particularity, and subtlety. By extending the transitive matrixial view to the planetary and cosmic sphere, we become witnesses to "a humanising factor of difference working in relation to objects and subject-matters, caring for humans and non-human animals, for the earth and for nature" (Ettinger 2019, 193), a factor which, in all its extent, has remained unrecognized until now. The concept of human subjectivity as formed by humans and entities *beyond* the human realm has thus far been indicated only marginally in the matrixial theory. Trans-spiration now offers a means to extend the carrying subjectivity to realms beyond human reference systems, thereby paving the path toward

Ettingerian environmental ethics. Opening up one's secluded "existential territory" to the drifts of the other – wherein the self acts as a carrier, "wholly supporting," in Levinas' words – extends beyond the human other. In each trans-spirational act, congealed corporeality transforms into a fluid, vibrating, restless substance with enough buoyancy to carry the incoming particles of the other. This poses questions regarding the permeability of the human entity and the expansion of the inner-physical and inner-psychic entity via a more significant, indeterminate external multiplicity. The boundary of individual "existential territory" becomes elusive, being in constant flux. Where does corporeality end, or what particles, air drifts, or (non-)Is does it encompass? Trans-spiration does not discriminate among the drifts approaching it, as the opening up always occurs from outside; the decision to accept certain drifts or to make distinctions for or against drifts is beyond the capacity of the self. Human conditions of assessment lose their significance; there is only the other. Such awareness accompanies a transformed perception of the environment and the world, rendering categorizations obsolete. Psychologically, corporeality could be described as a *psycho-planetary corpo-reality*, marked by the augmentation and mutation of anthropocentrism. Recognizing this expanded corporeality reveals another obligation of humanity: to allow oneself to be concerned not only by fellow human beings but also by the otherness surrounding them. Embracing this otherness dissolves the scheme of subject and object, inside and outside, without erasing their distinctions; even as the other is welcomed into the self, she/he/it remains foreign. The boundary between humans and the environment dissolves, as humans transition from clearly defined entities to a densely layered, agitated corpo-reality with an indeterminate multidimensional expansion: a body-space of mutual overlap and pervasion, a cosmos of constant drifts structured by open or closed membranes and carrying clouds of air. Celan eloquently captures this trans-spirational corporeality in his poem "Nachts."[1] In accepting the breath drifts of the other and the trans-spiritual unsettlement, humans' future and past manifest themselves. Traces lead back to the original aion from which respiratory rhythms and milieus once emerged, toward that which lies beyond time and beyond space. In trans-spirational breathing and the mutation of the bodily entity into a psycho-planetary corpo-reality, we engage in an unknown and unrecognizable encounter with God Himself, the source of breathing and being: "God, if we hold to this word, is the future itself, or rather the eternal reservoir beyond time and creating time, who constantly projects himself or pours himself into the present and who transforms the functioning of already created beings into sensible behavior and actions" (Ruyer and Walker 1980, 43). Nothing less than such outpouring occurs in every moment of trans-spiration.

Note

1 The poem has remained untranslated so far. The important passage translates as: "From the distant, from the blackened dream / Grove the breath drifts to us / And that which is lost goes around, large as the schemes of the future" (Celan 1952, 55).

8 In-Spiration

Letting myself be embraced by the breath drifts of the other, the unlocked individual encounters the primal airs breathed forth by the one who initially blew upon humanity, pouring out the spirit – the power and love of God Himself. Such an occurrence is inevitably tied to a specific moment in time: "And it shall come to pass in *the last days*, saith God, I will pour out of my Spirit upon all flesh: and your sons and your daughters shall prophesy, and your young men shall see visions, and your old men shall dream dreams" (Acts 2:17). In this biblical passage, the correlation between trans-spiration and the "last days" – the vacuum and trauma of the present moment – becomes evident: only when the "last days" have dawned and landscapes of desolation and scorched earth materialize, do paths unfold for the truly alive, the genuinely human(e). Subjectivity only reveals itself "after the 'death of the subject'" (Ettinger 2015a, 343); prophetic speech emerges only after the demise of language. As Paul Celan claimed, "language (...) had to pass through the thousand darknesses of deathbringing speech (...) Passed through and could come to light again, 'enriched' by all this" (Celan 2001, 395–396). Language that refuses to perish must confront the trauma of the present, the horrors, and atrocities of the "last days"; breathing that refuses to cease must begin to inhale the vacuum, the non-air – and through this metramorphic transformation discover renewed vigor and vitality. This attentive trans-spiration, as previously explored, possesses aesthetic attributes – yet it is not the artistry of bliss or anesthesia at work here, but rather the artistry of a "painter after the 'death of painting'" (Ettinger 2015a, 343): an artist after the demise of art. The requiem of the present harbors the potential for a new yet unknown language, for an expression of vitality beyond the realms of enclosure and reduction. In her theorization of this aesthetic procedure amid the "last days," Ettinger suggested that art serves as "a conductible sphere of emission, reception and transmission where waves and vibrations labour, where temporarily and partially strings and threads of spiritual consciousness are shared and transformed, then re-transmitted onwards, to the world" (Ettinger 2015a, 348). If unlocking for the other within trans-spirational breathing is inherently an aesthetic procedure – a process of floating-into-each-other – then this also implies that trans-spiration always grants access

DOI: 10.4324/9781003603672-12

to the incomprehensibility and "other otherness" of God. A spiritual consciousness of and with the transcendence of God is unfolded. Trans-spiration unveils a touch of the outpouring "of my Spirit upon all flesh," a participation not solely in the presence of the other but in the presence of the other Other – the grace and love of God. As Ettinger suggests, a spiritual occurrence unfolds in trans-spiration, an ephemeral unfolding of a "spiritual time-space-cluster" (Ettinger 2015a, 350), wherein the imprints of God are discernible in the breathing body, entwined and suspended within one another. Thus, it can be asserted that the biological-ethical act of trans-spiration evolves into a new form in which it becomes evident that "the human spirit is a particle of the spirit of the cosmos to which we are borderlinked" (Ettinger 2015a, 350) – that floating with the other is simultaneously always a floating with and in "the other Other" – a being touched by another divine-cosmic power, which – drawing from Celan's poem "Mit wechselndem Schlüssel" (see Chapter 7) – *provides the conditions* for the formation of snowflakes and their movement. In trans-spiration, the other enters "with a changing key" into my unlocked interior, into my floating and carrying space – a body of particles, flakes, waiting to welcome the other. A de-territorialization, a floating dance commences, an encounter where the other's word entwines with my flakes, floating-into-each-other. Yet, within this intertwining event, *something else* always lingers: a different drift, "the wind that carries you away," a force *causing* the snow to "gather around the word." What unfolds in this trans-spirational process is nothing short of the manifestation of infinity, an embodiment of the eternal breath of God. Trans-spiration establishes a connection "with the spirit of the cosmos to which we are borderlinked," momentarily tangible yet inherently intangible. This process, wherein "different body-psyche instances [connect] with one another and the human soul-spirit with the spirit of the cosmos," represents a "counter-traumatic human-consciousness vector" (Ettinger 2015a, 349). Trans-spiration transcends from a biological-ethical process to a *biological-ethical-transcendental event.* In becoming unlocked by the other ("you unlock the house"), in carrying the other within me ("drift with the flakes"), a *wind* materializes ("the wind that carries you away"), which, as Michael Welker articulates, "is both herald and medium of the process whereby God makes Godself present" (Welker 1994, 100). The correlation between wind and the manifestation of God finds vivid portrayal in an encounter between God and Elijah: "And, behold, the Lord passed by, and a great and strong wind rent the mountains, and brake in pieces the rocks before the Lord; but the Lord was not in the wind: and after the wind an earthquake; but the Lord was not in the earthquake: And after the earthquake a fire; but the Lord was not in the fire: and after the fire *a still small voice*" (1 Kings 19:11–12). It is noteworthy that the phrase "a still small voice" carries varied meanings: the Hebrew qôl (קוֹל) can be translated as "voice," but also encompasses "sound," "thunder," "cry," "rejoicing," "lamentation" – a term encompassing the expressions of humans, animals, objects, and ultimately of the divine. Particularly interesting is qôl's reference to the ethereal sound of flames,[1] the crackling of thorns,[2] the rustling of leaves,[3] and – of particular relevance to us – the sound of wings.[4] Thus, when such other breath wafts by at the moment of trans-spiration, we perceive the flutter of wings – a sound not heralding inaccessibility, punishment, or impending doom, but rather the *presence and embrace* of God, epitomized by the psalmist: "He will cover you with his feathers, and under his wings you will find refuge; his faithfulness will be your shield and rampart" (Ps 91:4). In every trans-spirational breath, therefore, unfolds a twofold movement: floating with the other who unlocks me, coupled with an affirmative embrace of this other and the readiness to carry her/his/its burden *and* becoming aware of a carrying breath that uplifts both me and the other – a breath way greater than the breath of me and the other. A sensation of being carried away emerges – potent yet gentle. Paul Celan's notion of "carrying away" must be understood not as rejection but as *elevation* into a vast expanse: "You have not given me into the hands of the enemy but have *set my feet in a spacious place*" (Ps 31:8). This elevation of "set my feet" can be construed as "to be appointed," as suggested in the Old Testament: "I said to them, 'The gates of Jerusalem are not to be opened until the sun is hot (...) Also appoint residents of Jerusalem as guards, some at their posts and some near their own houses'" (Neh 7:3). What becomes evident in this upward carrying is a simultaneous participation in the greatness, proximity, and inscrutability of God, a border-linkage to the "spirit of the cosmos," to the particles of divine transcendence. It is important to note that "the wind that carries you away" is no "strong wind," no "earthquake," nor a "fire" – not a grand, all-encompassing phenomenon, not a "hyperobject that is real yet inaccessible," as Timothy Morton posits.[5] Instead, this elevating wind transcends power, totality, and enclosing atmosphere. It is the small voice of rûah, a whisper, a flutter, a caressing. It is precisely within this gentle whisper that trans-spiration encounters *something beyond*, elevating the ethical-aesthetic act of trans-spiration into the transcendent realm. A breathing practice unfolds that surpasses totality, biologistic exclusivity, aesthetic sublimity, and the Kantian *noumenon*. Trans-spiration, in all its breadth, is no sublime process solely. It does not confine itself to aesthetics or ethics alone; it is always accompanied *by something else*, emanating from afar yet not necessarily distant. This transcendent form of trans-spiration is neither a *noumenal* operation nor the encounter with an inaccessible divine power and strength. Instead, it signifies a process wherein the divine infinity is indirectly perceived and experienced. Trans-spiration thus extends beyond the Kantian conceptual realm and its translations into the (post-)anthropocenic anesthesia.

*

In Kant's writings, *Noumenon* refers to what cannot be recognized by the human mind. This dimension lies outside the determinable and is manifested in Kant's further writings as the Sublime. For this *Noumenon,*

> we cannot apply any of our concepts of the understanding to it, this representation still remains empty for us, and serves for nothing but to designate the boundaries of our sensible cognition and leave open a space that we can fill up neither through possible experience nor through the pure understanding.
>
> (Kant 2000, 381)

In Kant's statement, what is shown is nothing less than the philosophical formulation of the opposition between the human "us" and the more-than-human "other," which is "a thing of which we can say neither that it is possible nor it is impossible, since we are acquainted with no sort of intuition other than our own sensible one and no other sort of concepts than the categories, neither of which, however, is suited to an extrasensible object" (380). We witness the emergence of a novel polarity between inside and outside, a demarcation between accessibility and inaccessibility. When Kant asserts the existence of phenomena that are "empty for us," pointing toward nothing but the sheer impossibility of apprehending them through any mode of understanding, it entails the inability to be moved or repelled by them – the potential inability of being carried away into an external space yet unknown; the incapacity to be touched by God. This also implies the impossibility of being affected by the other and the "other Other" – such that the Kantian subject remains settled in solitary respiration within an exclusively biologistic framework of breathing, thinking, and acting. If we hold on to the notion that God perpetually and solely remains distant, ineffable, potent, and threatening, then trans-spiration cannot achieve its full realization. However, as we learn from the biblical account, precisely "the Lord was not in the wind (...), the Lord was not in the earthquake (...), the Lord was not in the fire": God maintains a distance from any notion of totality, becoming revealed directly yet indirectly. Such a perspective on accessibility to an unknown and indefinite realm proves elusive in contemporary philosophical discourse. The current human-nature discourse postulates a global internal domain from which any intangible and incomprehensible externalities are fundamentally excluded – as we have seen in Sloterdijk's mystery of insulation. Referring to symbiotic paradigms of a flattened, harmoniously interwoven breath, the "outside" is either wholly negated and characterized as an undefined romantic idealization or rendered inaccessible. These tendencies toward internalization directly refer to the Kantian legacy, which prominently permeated 20th-century discourses in Freudian psychoanalysis, Heideggerian phenomenology, and the overarching narratives of modernist separation. In this context, the philosophical movement of *Object-Oriented Ontology* (OOO) can be read, of which one of its representatives says,

> there is a sense in which OOO could have existed immediately after Kant (...) If post-Kantian philosophy had taken this path instead of the German Idealist one, the last 200 years of philosophy would look entirely different. OOO can be seen as going back in time and making a course correction.
>
> (Harman 2017, 39)

With Graham Harman's OOO, the Kantian *Noumenon* expands to a novel dimension of thought that is incomprehensible for human knowledge and completely eludes "*any relation whatsoever*" (39). As Harman shows, the object of OOO cannot be revealed by any form of reductionism (neither "undermined" by physics, nor "overmined" by phenomenology), but always shows "a deeper reality not exhausted by my knowledge of it" (99). It is like the object of a "stone (...) unable to make direct contact with a cliff, a river, a sheet of paper, or the hand of a boy who throws it" (Harman 2017, 39). With the expansion of the Kantian *Noumenon* to OOO, *any comprehensibility* of reality is fundamentally rendered impossible from the beginning. This becomes apparent in the very statement that the "reality of the object [is] lying deeper than all human theoretical, perceptual, or practical activity" (Harman 2015, 99). OOO's frame of thought makes any relation to the world fundamentally impossible and remains caught in arbitrary fate. It is Timothy Morton's *Dark Ecology,* in which the dichotomy of OOO meets the current environmental discourse. With Morton, nature does not only remain a "problematic" (Kant 2000, 380) dimension but expands even further to become a "hyperobject" (Cf. Morton 2013), which "remind[s] us that *the local is in fact the uncanny*" (Morton 2016, 11). Morton's theory of nature illustrates that contemporary ecology is intertwining culture and nature in a peculiar cycle. Claiming that "the Anthropocene names two levels we usually think are distinct: geology and humanity" (7) means that androcenic creation transforms nature to such an extent that it must be regarded as a singular object, a *designed work of art* and an autonomous *place* in the sense of OOO – echoing Heidegger, who suggested that nature must be transferred via "clearing-away," "clear cut," and made "free from wilderness" (Heidegger 1973, 5) into cleared spaces, which he calls "places" (5). With Morton, nature precisely transforms into such a culturally designed place. As the Heideggerian interpretation of "place" suggests, place consists of the establishment of a condition that "hold[s] something free gathered around [it]" (7). Place is essentially about establishing an inaccessible and uncanny cleared space, where Morton identifies the paradox of the "hyperobject" – a condition which, despite its artificial androcenic design, increasingly eludes any human access and reference. In this way, Morton can state that "the Anthropocene is an antianthropocentric concept because it enables us to think about the human species not as an ontically given thing I can point to, but as a hyperobject that is real yet inaccessible" (Morton 2016, 25). Such inaccessibility of both humans and nature as an uncanny, violent force reflects the rigorous viewpoint of OOO. Interestingly however, this concept of the human and the more-than-human world remains strangely stuck in centuries-old Kantian polarities, only to shape them even more ossified into forms of an inaccessible total otherness.

*

In contrast to the radical alterity espoused in the works of Timothy Morton and Graham Harman, trans-spiration heralds a fundamental acknowledgment of an outside and the potential to *participate* in this exteriority: it expressly counters the assertion that any external realm "eludes any human access and reference." Indeed, in every genuine engagement with the world, a yet unknown unbounded space materializes; however, this precisely does not entail emergence within an inaccessible void, within a Heideggerian "clear-cut," within an "open center," but, as previously proposed through the psalmist, within a *space of liberty* wherein humanity is fundamentally situated and, in Celan's words, *carried away*: "You have not given me into the hands of the enemy but have *set my feet in a spacious place*" (Ps 31:8). What proves truly intriguing is that in trans-spiration, one can take a foothold within this spacious place, thereby standing "with one foot" amid the boundlessness of infinity and experiencing the transcendental breath of the spirit. The dimensions made accessible in trans-spiration are intimately linked to Levinas' conception of respiration: the lung function he proposes fulfills not only an ethical role but also a de-internalizing function wherein the subject is carried away from its interiority by meteorologies of infinite exteriority. Levinas writes: "For here there is a complex of significations deeper and broader than freedom, which freedom animates. Freedom is animation itself, breath, the breathing of outside air, where inwardness frees itself from itself, and is exposed to all the winds" (Levinas 1991, 180). In trans-spiration, one's corporeality is not only unlocked by the key of the other, initiating a floating-with-each-other – the subject also embarks on "the breathing of outside air," wherein subjective interiority "frees itself from itself" and from its mere trans-spirational trajectory. In the simultaneous "breathing of outside air," trans-spiration expands into in-spiration. The subject is situated within a vast, unexplored space; instead of invasive authority and inscrutable oppression, she/he encounters an unprecedented climate of liberty. In our examination of the meaning of *trans-spiration together with in-spiration* for the relationship between humans and nature, another biblical reference assumes significance: for the term "spacious place" in Psalm 31, the term merchab (מֶרְחָב) is employed, which, akin to the conventional translation, signifies "a broad or roomy place." In the book of Habakkuk, the same term is employed not merely for an indefinite expanse of liberty but is employed in a much more concrete sense. In Habakkuk 1:6, merchab interestingly connotes the "expanses of the earth," rendering the verse from Psalm 31 quoted earlier as: "You have not given me into the hands of the enemy but have set my feet *into the expanses of the earth.*" In floating-into-each-other, the subject experiences being carried by the wings of God into the expanses of the earth; it becomes a participant in this world. It is not merely a participant in interpersonal existence (trans-spiration) on one hand and divine transcendence (in-spiration) on the other, but also a participant in planetary immanence (*trans-in-spiration*). It is there that we fully understand the definition of subjectivity that Ettinger suggests: "the human spirit is a particle of the spirit of the cosmos to which we are borderlinked" (Ettinger 2015a, 350). The human being partakes in a tripartite alterity: in the *alterity of the human face*, the *alterity of divine infinity*, and the *alterity of planetary matter*. The subject is rendered threefold: existent, transcendent, and immanent; with human particles dispersed in all three directions. The subject no longer opposes these alterities, nor does it appropriate them; rather, it carries them and is carried by them; it elevates them and is elevated by them; it attracts them and is carried away by them. In trans-in-spiration, the human partakes in an unceasing respiratory movement, as suggested by Levinas: "It is the longest breath there is, spirit. Is man not the living being capable of the longest breath in inspiration, without a stopping point, and in expiration, without return?" (Levinas 1991, 182). In trans-in-spiration, the human aligns with the "longest breath" of rûah, the breath of God, which knows no end. This breath perpetually persists because it is directed toward "the expanses of the earth," and these expanses are vast, absolute, and extraordinary. Trans-in-spiration – as the tripartite breath movement of existence, transcendence, and immanence – opens the door to establishing a relationship with the alterity of face, infinity, and matter – one that is not afflicted by negative anxiety akin to a "hyperobject," but rather carries a productive, unfolding, and – in a manner similar to Deleuze and Guattari – *machinic* dynamism infused with awe, inner joy, and jubilant wonder. In this way, trans-in-spiration opposes the fundamental paradigm of anxiety characterizing worldly existence.

*

Such an anxiety-ridden reference to the world is prominently situated in Heidegger's concept of anxiety. According to Heidegger, "the essence of truth (...) subsists between clearing and concealing. It is the opposition of the original strife. The essence of truth is, in itself, the primal strife in which that open center is won" (Heidegger 1993a, 180). Truth is perpetually bound to the place of an inaccessible "open center," wherein truth itself is unearthed and contested in an ongoing dialectic struggle between "clearing and concealing." This "open center" of truth becomes associated with a specific condition – the condition of anxiety, representing the "basic attunement" (Heidegger 1996, 176) of Dasein at large. Anxiety and Dasein are thus intricately intertwined, as Dasein evades its potentiality in the "open center" of truth – there is "a flight of Da-sein from itself as an authentic potentiality for being itself" (172). In Heidegger's conception, this condition of anxiety mirrors Morton's later concept of the "hyperobject," being both intangible and all-encompassing – and fundamentally threatening: "the fact that what is threatening is nowhere characterizes what anxiety is about" (174). Heidegger's foundation of anxiety extends further still: it is the world as such: "that about which anxiety is anxious is being-in-the-world itself" (175). Hence, the source of anxiety is ultimately rooted in the world and in "being-in-the-world," pointing to its unpredictable and indefinite nature. This unveils a chain of references: Truth emerges in an intangible and thus to-be-escaped "open center" that human existence experiences as anxiety. This anxiety pertains to the intrinsic potentiality of the human subject as being-in-the-world itself, forging an immediate connection between anxiety and the world as such. Consequently, "being-in-the-world" and the world itself assume a decidedly negative connotation. Breathing for Heidegger, I argue, invariably entails an anxiety-laden experience wherein the linear trajectory of existence and its flight from the "open center" of truth, nothingness, and the world as such predominate. Specifically, for Heidegger, the world signifies absolute nothingness – a realm of space and time beyond any conceivable access or comprehension: "the Nothing is not 'something,' but nothing" (Heidegger 1993b, 45). In contrast to the Heideggerian nothingness we discern the biblical "spacious place," a dimension encompassing the "expanses of the earth." There is no anxiety but the movement of carrying away proposed by Celan – a becoming-elevated, an encounter with infinity within the face of the other, leading to the matter of the world. Humanity transcends its existing categories of being, becomes dislocated, and finds itself not as an outcast but, conversely, as an active participant in the world – not merely a spectator but a productive participant. Nevertheless, such dislocation *may* indeed be afflicted with a form of anxiety, rendering Heidegger's proposition not entirely unfounded; indeed, the notion that anxiety marks the "basic attunement" of Dasein as a fear of confronting its authentic potentiality holds some validity in Heidegger's argumentation; truly encountering oneself might be fearful. However, a crucial distinction lies in the fact that Heidegger's Dasein never indeed enters such an encounter but instead evades it. Ettinger now offers a novel perspective on Heidegger's conception of existential anxiety, reframing it in a manner that eschews avoidance of one's existential potential, not experiencing fear of entering the "open center," but instead embracing a caring, almost loving accessibility to the merchab – the boundless "expanses of the earth."

*

Ettinger proceeds from an etymological observation, namely the Hebrew word "'*de'aga*' with its verb '*lid'og*' (and its root '*d.a.g.*') that means 'to worry' and 'have concern for, to take care for and about'. I am concerned and involved as I take care, I care for your needs while I worry about you" (Ettinger 2011, 3). When Ettinger speaks of experiencing anxiety, or more precisely, being concerned about and being affected, there is always an altruistic component in this conception of anxiety: I am affected by, and, at the same time, I am concerned about the other. Anxiety is deprived here of its totalitarian connotation in that it is not threatening but rather a momentary agitation, swinging back and forth between wonder, concern, compassion, and reverence. The Heideggerian anxiety is differentiated here and means a "worry-caring" or "concerned arousal," consisting of "compassion and awe," which are "just as primary as anxiety" (3–4). This anxiety, nurtured by mutual concern and compassion, is not a threatening fear preventing the realization of one's potential but rather a catalyst for its full actualization. If anxiety is conceived as "'to worry' and 'have concern for, to take care for and about,'" then engagement in trans-in-spiration would also entail such matrixial anxiety. However, this distinct form of anxiety does not involve an existential plight but is always intertwined with "an option, an offer, a gift" (7). In this context, anxiety marks the opportunity to engage with and embrace the other's alterity affirmatively. However, the actual embrace of this gift necessitates an ethical decision, a tribute to the other. Thus, "worry-caring" encompasses a dialogical condition – an invitation for immersing with/in the other, a readiness to be carried away into the "expanses of the earth." The other ceases to be an ominous force, and the world is no longer a hostile realm threatening human existence. Instead, dimensions of reciprocal "worry-caring" emerge among the three alterities of existence, transcendence, and immanence; in trans-in-spiration, the animate and the inanimate float-into-each-other, perpetually demanding renegotiation. The other may be the cause of the "uncanny awe" (8), but never the total inaccessible otherness, since it also always creates the gift for an active, mutual "affectual attuning in arousal" (8). The other in this "wonder-worrying" process transforms into a "feminine almost-Otherness and *side-by-side-ness*: a *besidedness* that permits to think co-emergence of I *and* non-I. The phallic knot composed of sacrifice, death and the feminine dissolves in favor of the co-incidence of death and life, sacrifice and solace with-in the feminine" (Ettinger 2006c, 101–102). Ettinger proposes a co-existence in which one's subjective boundaries open up and become permeable to an interplay between the I and the other as non-I. This interplay unfolds a temporal continuum in which the face, infinity, and matter are interwoven and manifest as particles of the I, as "(non-)I grains." This continuum does not exist from the beginning, and it is not a given imaginary or symbolic framework whose contents would merely have to be filled in. Instead, it is a process that has to be constantly acquired and negotiated anew, without any guarantee of achieving it at any time. A choice is always necessary. In this process of negotiation and dissolution of identity and subjective boundaries, the "almost-Other" (Ettinger 2006b, 195) is *never inaccessible,* and her/his/its affirmative acknowledgment presents her/him/it as an intrinsic yet unknown part of one's self. In this state of "besidedness" within trans-in-spiration, the foundational disposition of anxiety reveals itself as a "concerned arousal," as the unfolding of a gift and possibility: to be moved by the face of the other, to be elevated into the infinity of God, and to venture into the vast "expanses of the earth." A *different consciousness* emerges, aware that each breath taken is not a solitary, isolated event but rather part of a relational continuum. In trans-in-spiration, a dialogical, or rather: tria-logical interchange unfolds among the three realms of alterity – face, infinity, and matter – which are never wholly foreign but continually manifest as the non-I particles of the I. The human subject cannot be split from the world: opposite to Heidegger's Dasein standing in anxious relation to the world and avoiding it, the Ettingerian feminine-matrixial subject is positioned in an attentive, caring, curious, affirmative, and open stance. With every breath, the subject acknowledges participation in an unceasing and trans-connected breath, perpetually engaged in a "feminine borderswerving, borderlinking, and borderspacing" (Ettinger 2016a, 153). This implies a condition of trans-subjective transformation, a continuous embracing, hovering, and participating in the realms of existence, transcendence, and immanence. The human being is no longer a self-contained entity; the anxiety of the world vanishes as one becomes trans-in-spired, experiencing that the human always entails and carries the more-than-human. Through openness to the other, through drifting with the other's flakes, becoming carried away into infinity, descending into the earth's expanses – one is human yet transcending the human realm. Both here and there, one transpires and becomes inspired, joining the eternal breath of the divine and the cosmic breath of creation. The time granted to humans is nothing short of a continual trans-in-spiration by the Spirit of God – summarized succinctly by Levinas in a conversation with Ettinger: "time is the breath of the spirit" (Ettinger 1993, 21).

Notes

1 "Like the noise of chariots on the tops of mountains shall they leap, like the noise of a flame of fire that devoureth the stubble, as a strong people set in battle array" (Joe 2:5).
2 "For as the crackling of thorns under a pot, so is the laughter of the fool: this also is vanity" (Ecc 7:6).
3 "And the sound of a shaken leaf shall chase them; and they shall flee, as fleeing from a sword; and they shall fall when none pursueth" (Lev 26:36).
4 "And when they went, I heard the noise of their wings, like the noise of great waters, as the voice of the Almighty, the voice of speech, as the noise of an host: when they stood, they let down their wings" (Eze 1:24), "I heard also the noise of the wings of the living creatures that touched one another, and the noise of the wheels over against them, and a noise of a great rushing" (Eze 3:13), "And the sound of the cherubims' wings was heard even to the outer court, as the voice of the Almighty God when he speaketh" (Eze 10:5).
5 Reading Timothy Morton's *Hyperobjects: Philosophy and Ecology after the End of the World* (2013), Clayton Crockett claims: "Religion, understood in Morton's terms, is a massive, complex, and distributed hyperobject. *And God, too, is a hyperobject.* God does not simply exist in the way that most objects do, or even most hyperobjects that Morton describes" (Crockett 2018, 87).

9 Metreorology

In Ettinger's concept of "feminine borderswerving, borderlinking, and borderspacing" (Ettinger 2016a, 153), the breathing subject engages in a tripartite relationship with the world: resonating with other humans, with the infinity of God, and with the abundance of planetary matter. The subject transcends the role of an observer who would objectify the other, God, and the earth. Instead, each completed breath of trans-in-spiration reveals the placement of the self within the other: the other as I, not identical to the I, but rather as non-I – a particle that relates to me, concerns me, and of which I am composed and recomposed with each breath. By opening oneself to the other, by allowing the other to open oneself, one welcomes and embraces the gift offered by the other. The I begins to fragilize, opens its boundaries, and becomes porous and permeable. It breathes into a dimension where the direction of its breath cannot be predetermined; there is only the undefined certainty of the potential touch by the other, the divine infinity, and the earth. This situates the individual within an all-encompassing cosmic context that surpasses Heidegger's fourfold. Recalling Heidegger's arrangement of Dasein within the knot of "earth and sky, divinities and mortals" (Heidegger 2012, 18), which Ettinger expands into a *threshold* where earthly and cosmological forces intertwine closely, we notice a further step taken by "feminine borderswerving, borderlinking, and borderspacing." Here, no rigid knotting occurs, as for every knot tied, one authority would be required to bind it. In the Ettingerian matrixial theory, there is no organizing authority dictating human fate. With its exclusive focus, the orientation toward fate evident in Heideggerian anxiety upholds a consistently phallocratic order. In the matrixial theory, such an order is abolished; there is no singular fate, only reciprocal offerings and revelations of possibilities. The other enters, lifts, and situates the self in a spacious place; these movements occur not centrally but in reciprocity and resonance. The planetary cosmos unfolds as a medium of mutual addressing, welcoming, engaging, uplifting, and carrying away. The vibrating medium, wherein every trans-inspirational breath occurs, may be termed *air* – but only if air is understood etymologically as *desire, need, admiration, praise, or love*. Such derivation from the Proto-Indo-European root *lewb^{h}- (to desire, covet, admire, praise, love) indicates air as a fundamental medium of non-sexual eros – where I

DOI: 10.4324/9781003603672-13

never remain a stranger but recognize myself in the other. I am in the other as the other is in me – signifying mutual responsibility. Air becomes the medium of mutual praise, admiration, desire, and love. Air makes possible what Max Scheler, with a criticism of Heidegger's anxiety, proclaimed as a basic condition of life: "The first turn to the world-immanent is Eros not repulsion, anxiety, or flight (...) What unlocks the world to us is 'love' and not anxiety" (Scheler 1976, 294). In this loving medium of air, rigid and one-sided identities transform into permeable ones. In the air, I become aware of becoming a tree just as the tree becomes aware of becoming human. Yet, clear identity attributions of humans and trees dissolve, as there are only particles, drifts, or momentary *condensations* that lovingly shift with each trans-in-spirational movement. The characterization of air as a medium of love and desire, of affective attraction and repulsion, of opening and being lifted, is crucial for concretizing a trans-in-spirational breathing practice and its implications for a relation to the world – facets often overlooked in the hegemonic definition of air.

*

During his expeditions, Alexander von Humboldt made a profound observation that air carries

> neben den entwickelten Geschöpfen (...) auch zahllose Keime künftiger Bildungen, Insecten-Eier und Eier der Pflanzen, die durch Haar- und Feder-Kronen zur langen Herbstreise geschikt sind. Selbst den belebenden Staub, den, bei getrennten Geschlechtern, die männlichen Blüthen ausstreuen, tragen Winde und geflügelte Insecten über Meer und Land den einsamen weiblichen zu. Wohin der Blik des Naturforschers dringt, ist Leben, oder Keim zum Leben, verbreitet.
>
> (Humboldt 1806, 4–5)[1]

Humboldt conceptualized the air as a unitary medium encompassing diverse living identities, describing it as a "Luftmeer, in das wir getaucht sind, und über dessen Oberfläche wir uns nicht zu erheben vermögen" (5).[2] Within this ocean, entities interact, fertilize, repel, and continue their airborne voyages. Humboldt's perspective aligns with Aristotle's concept of meteorology, which examines singularities suspended in the air – hence the term "meteorology" derived from μετέωρος, meaning "floating." In his *Meteorologica* (around 340 BC), Aristotle writes:

> It remains to consider a subdivision of the present inquiry which all our predecessors have called *Meteorology*. Its province is everything which happens naturally, but with a regularity less than that of the primary element of material things, and *which takes place in the region which borders most nearly on the movements of the stars*. For instance the

> milky way, comets, shooting stars and meteors, all phenomena that may be regarded as common to air and water, and the various kinds and parts of the earth and their characteristics.
>
> (Aristotle 1952, 5)

Aristotle delineates meteorology as the study of singularities occurring naturally but with less regularity than primary material elements, spanning from the earth's surface to the celestial realm. This conception encompasses various atmospheric occurrences such as clouds, wind, rain, and thunderstorms, illustrated in modern weather reports' satellite imagery. Both Aristotle and Humboldt define air as the sphere of the interaction of self-contained singularities with an impenetrable shell. These entities inhabit an ocean – reminiscent of Aristotle's notion of *ether*, the fifth element – where they engage in causal networks. Even when these interactions enhance in complexity, such as unpredictable weather phenomena, they adhere to defined physical patterns. This also remains the case when such interactions take on symbiotic, holistic dynamics, when the world is understood

> as a knowable entity – a single, interconnected whole – but in a sense that lacked the secure stasis of maps, parlor globes, or pre-Darwinian cosmologies. Instead, it meant grasping the planet as a dynamic system: intricately interconnected, articulated, evolving, but ultimately fragile and vulnerable. Network, rather than hierarchy; complex, interlocking feedbacks, rather than central control; ecology, rather than resource.
>
> (Edwards 2010, 1–2)[3]

Such understanding of air and meteorology as interconnected feedback loops operates within known and established frameworks, devoid of love's world-determining force, as Scheler suggests. Rather, it remains within the boundaries of mechanistic dynamics that do not allow for any kind of being touched or carried away; a mechanistically understood air claims a clearly delineated atmosphere, full of enclosing tendencies, even if such tendencies are hidden under the guise of a blissful symbiosis. It is noteworthy that the term "ecology" serves as an umbrella term for such a mechanical worldview.

*

"Ecology" permeates the (post-)anthropocenic discourse to such an extent that one could argue for a downright "ecologization" of thought; each contemporary practice is assessed through the lens of ecology and ecological awareness. Yet, amid profound geological and geopolitical shifts, this notion of ecology is far from positive; rather, it aligns with Heidegger's paradigm of anxiety, portraying it as an unavoidable constant of human existence to be escaped from. Slavoj Žižek remarks that "the by far predominant version of ecology is the ecology of fear, fear of a catastrophe (...) This ecology of fear has all the chances of developing into the predominant form of ideology of global capitalism, a new opium for the masses" (Žižek 2007).[4] This statement underscores that the "ecology of fear" is not merely a negative element but holds significant sway over human thought and action. When ecology, breathing, and existing in the world are perceived as inherently fearful, it not only obstructs access to the seemingly threatening otherness but is also actively *desired* by humans. This raises a crucial question: How has a fate-oriented, negative mode of thinking that assigns passivity to humans – a perspective observed in Freud, Heidegger, Kant, and even Aristotle – exerted such profound influence over centuries, even millennia? This influence cannot solely be attributed to the prominence of these thinkers or their spheres of influence; rather, it stems from the unconditional acceptance of these fate-centered paradigms by humanity itself. Here, we revisit a foundational point where Lyotard stated that "the English unemployed did not become workers to survive, they (...) *enjoyed* the hysterical, masochistic, whatever exhaustion it was of hanging on in the mines ..." (Lyotard 1993, 111). Such inclination to enjoy elucidates the opioid effect of the "ecology of fear" for the great number of humans who prefer to remain ensnared in a world dictated by fate rather than exert the effort to break free from this cycle of mechanistic symbiosis governed by cause and effect. Wilhelm Reich echoes a similar view in his critique of reductionist Freudian psychoanalysis, positing that human existence inclines toward irresponsibility, immobility, and character armoring, hindering any form of commitment, openness, or ethical-aesthetic action (Cf. Reich 1953). To avoid succumbing to the narratives of the "ecology of fear" and to escape confinement within a closed and armored societal ocean, it becomes imperative to reimagine the concept of air and challenge the hegemonic framework of Aristotelian meteorology. The foundational characteristics of air, originally conceived as a space of love, care, sharing, and resonance, are inherently marked by its *aesthetic (perceptive) quality*. Trans-in-spirative respiration begins with an aesthetic breath – a *perception* of the other – even before direct confrontation occurs. Without this primordial aesthetic orientation, trans-in-spirational breathing remains stagnant, mechanical, and purely biologistic, with the other remaining distant and inaccessible, dominated by physics over aesthetics. Deciding for aesthetic breath entails an acknowledgment that one does not breathe alone, passively resigning to fate, but instead recognizes the other's offering – a possibility, a gift – granting the opportunity to perceive the self in the other and to transcend prevailing egocentrism. Franz Schubert's (1797–1828) song "An die Musik" ("Ode to Music," D. 547) exemplifies how the aesthetic process of trans-in-spiration unfolds.[5] The lyrics by Schubert's contemporary Franz von Schober (1796–1882) elevate the transformative power of art and aesthetic agency. It praises the gentleness and grace of art ("du holde Kunst"), portraying it as a force not universal but intimate and accessible only to those in need. Art possesses the potential to unsettle, allowing those who engage with it to perceive alterities that can transport them into entirely different realms. The line "when the savage ring of life entwines me" opens the song; this line appears significant to us in two respects: art evidently can break free from brutal mechanized cycles. Such cycles may indeed be akin to those of Aristotelian meteorology, re-encountered scientifically in Margulis or Timothy Morton, and philosophically in Heidegger or Freud, now consolidated in the (post-)anthropocenic ecology of fear and the predominance of anesthesia. Such a "savage ring of life" would also entail living in complete passivity and negation of one's own capacity for assuming responsibility; a life that does not wish to or cannot perceive the gift of otherness. On the other hand, this invocation of art signifies a departure from entwined constellations, as elucidated in Heidegger's "fourfold" or in the contemporary discourse of "entanglements." The prominence of the concept of entanglement does not necessarily point to Ettinger's "feminine borderswerving, borderlinking, and borderspacing," but rather suggests a new form of contemporary anesthesia: it proclaims that the world is no longer flat, synchronous, or painless but intricately "entangled" in manifold ways. Entanglement can thus be understood as a new heading above the hegemonic ecology of fear: ecology is entangled to such an extent that intervening in these entanglements is no longer possible; openness, receptivity, and fragility are made impossible. Entanglements prevent taking part in one another, being carried away by each other. Being touched by the infinity of God is no entanglement but an event; not physics but transcendence; not identity but porosity; not fertilization but floating-into-each-other. From this awareness, Ettinger decidedly opposes the definition of human procreation as solely based on the biological function of "the coalescence of two germ-cells which are differentiated in a particular way" (Freud 1955, 44). She asserts that

> the matrixial Eros marks a different kind of life-drive then, for which the language of germ cells and the idea of the principle of life's longing for its inanimate and even inorganic origin is unfit, a different kind of life-drive that relates to the transgenerational as well as the horizontal transmissibility of traces of memory and oblivion to what was affected in/by the individual between birth and death.
>
> (Ettinger 2019, 190)

A dimension of air understood as loving, caring, and being concerned, thus, is not a biological ocean of fusion and mutual reproduction; we are encountering a dimension that is effective beyond the determinable, beyond the positivistically demonstrable: a meteorology beyond the duality of the life and the death drive.

*

The concept of the death drive (or death instinct), which first appeared in Freud's *Beyond the Pleasure Principle*, was initially based on the observation that the human subject exclusively follows "the tendency towards stability" (Freud 1955, 9), i.e., the quickest dissolution of all psychic tensions. The mechanisms of this dissolution Freud referred to as the *pleasure principle*, which "strive towards gaining pleasure; psychological activity draws back from any event which might arouse unpleasure. (Here we have repression.) Our dreams at night and our waking tendency to tear ourselves away from distressing impressions are remnants of the dominance of this principle and proofs of its power" (Freud 1958, 219). Freud recognized, however, that the intended psychic homeostasis of the subject usually remains absent, so that the subject is forced "to form a conception of the real circumstances in the external world and to endeavour to make a real alteration in them" (219). Freud named this concept the *reality principle*. In confronting everyday reality, this principle allows the subject to take detours on the way to fulfilling the pleasure principle and to continually modify one's striving to avoid unpleasure. Formulated in 1911, Freud later expanded the pleasure principle, as disturbed by the reality principle, with one of his most important concepts – the *death instinct*. The death instinct marks the "beyond" of the pleasure principle. Although Freud stated, "that there exists in the mind a strong *tendency* towards the pleasure principle [via the reality principle]," (Freud 1955, 9) this condition is "opposed by certain other forces or circumstances, so that the outcome cannot always be in harmony with the tendency towards pleasure" (9–10). For Freud, these resisting forces were those of the death instinct that define the orientation of every psychic and physical life. While the death instinct, at first, relates to the pleasure principle, it also goes beyond it in its scope. Referring to human biological development, Freud initially stated that

> the attributes of life were at some time evoked in inanimate matter by the action of a force of whose nature we can form no conception (…) The tension which then arose in what had hitherto been an inanimate substance endeavoured to cancel itself out. In this way the first instinct came into being: the instinct to return to the inanimate state.
>
> (38)

In the Freudian death instinct we see a consistently determined life cycle that presupposes a small number of predetermined principles in order to ensure the vitality of life. Freud attributed the agency of the death instinct not only to humans but to every organism – "all living beings" (39) – and drew an essential conclusion in identifying the performing phenomena of the death instinct:

> It was not easy, however, to demonstrate the activities of this supposed death instinct (…) It might be assumed that the death instinct operated silently within the organism towards its dissolution, but that, of course, was no proof. A more fruitful idea was that a portion of the instinct is

> diverted towards the external world and comes to light as an instinct of *aggressiveness and destructiveness.*
>
> (Freud 1961, 119)

Here, Freud equated the death instinct with a destructive and aggressive instinct toward the world and contrasted this male-connoted drive (Tanathos) to its counterpart, the (female) life or sexual instinct (Eros).[6] Although Freud remarked that both opposing instincts – death and life instinct, Eros and Tanathos – are, in fact, involved in the formation of subjectivity and the psychic apparatus, for him, the (male) death instinct was the one that ultimately prevails. This observation becomes clear as Freud considered the death instinct to be the "original course" (Freud 1955, 38) for humans. In contrast, the life instinct is somewhat related to later developed cultural ambitions. Thus, a temporal hierarchy in the two instincts can be seen, as well as the two polarities of male strength and activity, and female weakness and passivity. When Freud expressed in the formulation of the castration complex that femininity also grounds on a tendency toward masochism (Cf. Freud 1964), Freud's opposite equation of the death instinct with (male) sadism is obvious. "It is in sadism, where the death instinct twists the erotic aim in its own sense and yet at the same time fully satisfies the erotic urge, that we succeed in obtaining the clearest insight into its nature and its relation to Eros" (Freud 1961, 121). The Freudian depiction uncovers a perpetual struggle between two drives, entwining the human subject within an entirely biologistic and quasi-archaic fate. The polarity of Eros and Thanatos depicts a cyclical trajectory of human vitality, devoid of an inclination toward mutual floating-into-each-other and receptivity of the other. Instead, it manifests as a cycle marked by dynamics of male aggression, repression, and subjugation. The Eros identified by Freud within the life instinct lacks the attributes of love, care, vulnerability, and perceptiveness that we associate with our matrixial formulation of air.

*

Here we again find the importance of the "matrixial Eros," wherein "subjectivity in itself transgresses the individual subject. Relation-without-relating leads the subject toward responsibility on the unconscious level of partiality and transgressivity treasured upon traces of the archaic co-implication and co-affection" (Ettinger 2006c, 105). A realm of "partiality and transgressivity" unfolds, signifying the presence of an Eros beyond mechanistic principles; an Eros that is not oppressive or submissive, but structured in an entirely different way: within it occurs "a kind of 'quantum-like' resonance of I/non-I elements (…) to reach some awareness of that level of transconnectedness between different elements in between me and the other, me and the cosmos, beyond the threshold of our sense-perception" (Kaiser and Thiele 2018, 117), or, in other words: a "subjective resonance via the aerials of the psyche, this transjective, quantic-entanglement-like resonance between the I and non-I as becoming-subjects" (Ettinger 2014, 125). Such an Eros, effective in a cosmic interplay between the face of the other, the infinity of God, and the expanses of the earth, can no longer be situated within a meteorological dimension. Instead, it is necessary to transition the mechanistic procedures of meteorology into a matrixial *metreorology*. Metreorology, thus, cannot be described merely as the science of predetermined and interacting entities in the air, but as the *science of breathing for the other*, the realization of being carried away and set free in trans-in-spiration. *In metreorology, air is traced back to its true meaning as love, desire, the recognition of the other.* Humans find themselves situated in a medium that enables immediate fragilization at any time, continuously offering itself as a gift. However, trans-inspiration is never accompanied by loud and sounding phenomena but by "a small voice": not by a mechanical voice of attraction and repulsion, not by the battle-roar of Eros and Tanathos, not by the incessantly repeating litanies of flatness and bliss. *The metreorological dimension is a dimension of silence, of emptiness, of non-air, of mute scorched earth*:

> Matrixiality does not oppose solitude and aloneness. Often, even the contrary is true: listening to that which is not "you" and which doesn't reflect your identity demands silence, slowing-down, blind gazing, awe, respect, compassion; allowing wondering, tolerating anxiety; welcoming the contingent and the unknown.
>
> (Kaiser and Thiele 2018, 117)

Cioran suggests a similar "mute paradigm" for true contact between the I and the (non-)I, stating: "True contact between beings is established only by mute presence, by apparent non-communication, by that mysterious and wordless exchange which resembles inward prayer" (Cioran 1986, 7). Thus, air characterizes itself as a medium where the silent drifts of rûah flow, where the invocations of the other reach me like prayers and the hopeful gaze of the other's face looks toward me – air as a medium where "deësis" (δέησις) can be immediately experienced and realized.

*

Let us revisit Schubert's *Ode to Music* once more: here, art is not merely praised as a means to transcend into a superior realm. Primarily, every aesthetic engagement with the world – initiated first and foremost through acts of trans-in-spiration – kindles "warm love in my heart." Each aesthetically perceived breath, as it floats into the realms of existence, transcendence, and immanence, brings forth the air in its most *original condition of kindling the heart with warm love*, resonating within a space beyond mere mechanical appropriation and superficial entanglement. A realization of an overlapping identity unfolds, unveiling the practice of "caring and carrying and encountering the other's vulnerability, then vulnerability and care, carrying and self-fragilization" (Kaiser and Thiele 2018, 117). Such realization emerges from an initial sigh ("often a sigh has escaped from your harp"), akin to a prayer-like breath ("a sweet, sacred harmony of yours") that emanates from the other. This breath always accompanies the imperative decision to accept it, unlocking one's gates for the other to enter with her/his/its key. Embracing the other in this manner initiates the fragilization of one's own identity and sets forth a mutual floating-into-each-other, paving the way to "open[ing] up the heavens to better times for me." Metreorology, therefore, primarily signifies the *science of sighs*, operating on various levels: First, on the level of the sighs of fellow human beings present to me, from which every trans-in-spiration originates; second, on the level of the small voice of God; lastly, on the level of a breath that carries me away and sets my feet into the freedom of vast planetary space – from which humans have to inevitably return to their everyday realities again and again. Trans-in-spiration is not a perpetual phenomenon; it occurs fleetingly, in the briefest of moments. Unsettlements and restlessness are ephemeral; humans are consistently drawn back from the presence of metreorological trans-in-spiration into the mechanized realm of objective reality. Perceiving the world as a Levinasian "Vous" and welcoming the otherness of the other necessitates a recurring shift in perspective, regarding the world as an object again. This entails an ongoing transformation, eluding mechanistic cause-and-effect relationships. Trans-in-spirational breathing does not arise from predetermined causality but from a momentary, unpredictable *decision for the other*. Yet, it must always depart from this other again, returning from the celestial wings of God and the floating-with the world to the harsh reality of mechanics and the constraints of linear time. As Martin Buber articulates:

> The man to whom freedom is guaranteed does not feel oppressed by causality. He knows that his mortal life is by its very nature an oscillation between You and It, and he senses the meaning of this. It suffices him that again and again he may set foot on the threshold of the sanctuary in which he could never tarry. Indeed, *having to leave it again again* is for him an intimate part of the meaning and destiny of this life. There, on the threshold, the response, the spirit is kindled in him again and again; here, in the unholy and indigent land the spark has to prove itself.
>
> (Buber 1970, 101–102)

*

The (post-)anthropocenic reality continually presents *the opportunity* to perceive the sigh of the other, to expose oneself to it, to welcome it, to allow oneself to be ignited by it, and to float over this vast space of the earth in communion with the infinity of God. In trans-in-spirational breathing, the confines of space and time dissolve, facilitating a transition from trans-subject to trans-ject, where the dissolution of the "I," the "self," and even the biological-bodily unity occurs, *leading to only momentary quantum-like corpo-real compositions*. These compositions arise from the floating-with I and non-I(s), carried away into an archaic past and an indeterminate future. Subjects, therefore, can be discerned as essential *dust*. It is not solely a matter of "for dust thou art, and unto dust shalt thou return" (Gen 3:19), but rather that the human destiny is intrinsically linked to being dust. True vitality emerges only when humans recognize themselves as dust and embrace the prospect of breathing aesthetically *as dust-grains*. In this particular understanding of existence, humans are in a perpetual state of indefinite transition; they never remain stagnant, being continually committed to the other beyond suppression or appropriation. Ettinger claims: "Something of *it* will arrive, to me and to-from the cosmos. You are, I am, grains of dust in the eyes of the vast glowing vault" (Ettinger 2015b, 354). From the world of the "it" and a broader, indeterminate cosmic context, something reaches me; the I and the It particulate, becoming grains of dust engaged in a trustful relationship, inscribing themselves upon each other and carrying themselves to new shores.

*

An ethical comprehension of the world rooted in the Ettingerian matrixial theory can only be fully grasped within the realms of indeterminacy and floating-with/for/toward each other. Confronted with the entwining reality of the (Post-)Anthropocene, which fundamentally originated from phallocratic meteorologies, it becomes imperative to perceive air – the medium that sustains all life – from a matrixial viewpoint, as a medium of love and trans-in-spiration. Here, the other drifts toward me, and I recognize myself within her/him/it, assuming responsibility as an integral part of myself. Viewing air as a medium of love and desire imbues it with a distinctly feminine essence, elucidated further by metreorology. In his 1883 poem "The Blessed Virgin Compared to the Air We Breathe," Gerard Manley Hopkins eloquently captures this matrixial essence of air: "Wild air, world-mothering air, / Nestling me everywhere" (Hopkins 2002, 158). Once air is perceived as a nurturing and maternal force, characterized by its capacity to "nestle" and "nurse," then the related metreorology becomes the foundation of an Ettingerian environmental ethics. In this framework, I find myself in silent floating communion with my neighbor, with God, and with the stone beside me. The (Post-) Anthropocene, with its pervasive ecology of fear and anesthesia, empowers me to consider air anew, precisely *in facing* the profound traumatic changes that the planetary environment undergoes day by day. In doing so, I no longer perceive air solely through the lens of reality, as defined by the Freudian reality principle, but rather as a realm of possibility:

> "So the sense of possibility could be defined outright as the ability *to conceive of everything there might be just as well*, and to attach no

> more importance to what is than to what is not. The consequences of so creative a disposition can be remarkable, and may, regrettably, often make what people admire seem wrong, and what is taboo permissible, or, also, make both a matter of indifference (...) But the possible includes not only the fantasies of people with weak nerves but also the as yet unawakened intentions of God. A possible experience or truth is not the same as an actual experience or truth minus its "reality value" but has – according to its partisans, at least – *something quite divine* about it, a fire, a soaring, a readiness to build and a conscious utopianism that does not shrink from reality but sees it as a project, something yet to be invented. After all, the earth is not that old, and was apparently never so ready as now to give birth to its full potential.
>
> (Musil 1995, 11)

In metreorology, a revelation of the divine, the utopian, and something yet undiscovered emerges, offering a novel relationship with the world – one that awaits consideration. The mere task lies in unlocking one's own door.

Notes

1 Translated as: "Besides the developed creatures (...) also countless seeds of future formations, insect eggs and eggs of plants, which are protected by hair and feather crowns for the long autumn journey. Even the invigorating dust that the male flowers scatter when the sexes are separated is carried by winds and winged insects over sea and land to the solitary female flowers. Wherever the naturalist's gaze wanders, life, or the seed of life, is spread."

2 Translated as: "Ocean of air into which we are immersed and above whose surface we are unable to rise."

While Alexander von Humboldt famously described the atmosphere as a unitary medium of diverse living identities, the concept of "Luftmeer" itself predates his formulation. Evangelista Torricelli's 17th-century experiments with barometric pressure and Immanuel Kant's discussions of the atmosphere in *Physical Geography* likewise invoke the idea of air as an oceanic expanse. Taken together, these earlier sources illustrate a longstanding philosophical and scientific tradition that conceives of air not merely as an invisible background but as a life-enabling envelope.

3 As Paul Edwards (2010) makes clear in *A Vast Machine*, the conceptualization of weather and climate as interconnected feedback loops has always involved a sense of almost Kantian sublimity in facing the inherent complexity and partial unpredictability of atmospheric systems – an attitude that dates back at least to John Ruskin's weather observations in the 19th century.

4 Mike Davis coined the term "ecology of fear" in his 1998 book of the same name. He portrays Los Angeles as a milieu in which the deliberate creation and control of fear in the form of human-made/man-made natural disasters, such as "total fire suppression," marks the daily political order – both on the part of the political leaders and on the part of the citizens.

5 The lyrics are: "Du holde Kunst, in wieviel grauen Stunden, / Wo mich des Lebens wilder Kreis umstrickt, / Hast du mein Herz zu warmer Lieb' entzunden, / Hast mich in eine bess're Welt entrückt! / Oft hat ein Seufzer, deiner Harf' entflossen, / Ein süßer, heiliger Akkord von dir / Den Himmel bess'rer Zeiten mir erschlossen, /

Du holde Kunst, ich danke dir dafür!" translated as: "O blessed art, how often in dark hours, when the savage ring of life entwines me, have you kindled warm love in my heart, have transported me to a better world! Often a sigh has escaped from your harp, a sweet, sacred harmony of yours has opened up the heavens to better times for me, O blessed art, I thank you for that!"

6 In his writings, Freud speaks of Eros and the death instinct. The denomination of the death drive with the Greek god of death (Tanathos) is often invoked, but not used by Freud himself.

Conclusion

Reflecting on these implications of matrixial breath and the effects of an aesthetic trans-in-spirational relation to the world is overshadowed by terrible news from my hometown, Graz, Austria: On the night of March 23, 2024, a homeless man sleeping in a house entry was doused with a flammable liquid and set on fire by an unknown person. During the investigation, the perpetrator, a homeless man by himself, admitted that he had lit a fire in this place to warm himself but *had not perceived* the man sleeping there. In all its tragedy, such argumentation reflects the anesthesia of the androcenic present, in which humans are no longer sensitized in any way to the otherness of their fellow human beings and the world around them: humans light a fire – without realizing that they are setting another person on fire. The incident manifests the hegemony of scientific meteorologies, wherein self-contained entities act according to physical principles without perceiving or considering the other; it seems that humanity is replaced by thermodynamics. As this book argues, transferring principles and structures of physics to human thought and action can implicate unforeseeable dynamics: meteorology, beyond the discipline of weather and climate phenomena, can also take on the discipline of mutual destruction. Disregarding the other is what matrixial breath stands up against most fundamentally; as trans-in-spiration suggests, in carrying the particles of I and (non-)I, scientifically defined identities of self and otherness are lifted in favor of a healing and truly humane de-territorialization. At the same time, such a view also means that opening for the other is not associated with Heideggerian anxiety about one's existence. Welcoming the other happens without Abrahamic sacrifice.

*

Trans-in-spirational processes introduce the formulation of *Ettingerian environmental ethics*, which suggest a counter-position to ethical holism. In recent decades, various positions of environmental ethics have developed, which can be listed according to their scope of influence: anthropocentrism (Dierks et al., 2016, 11), pathocentrism (Reder et al., 2019, 25), biocentrism, ecocentrism (Reder et al., 2019, 29), and holism (Teutsch, 1985, 46). While anthropocentrism sees the relationship with the environment primarily in

DOI: 10.4324/9781003603672-14

terms of its usefulness to humans, and pathocentrism, biocentrism, and ecocentrism claim a decentration of human agency in favor of the ability for suffering of living beings, plants, and even inanimate matter, holism proposes a coherent, systemic, and unitary position in the relationship between humans and nature. This holistic view serves to clarify the positions of Ettingerian environmental ethics. We have already seen that matrixial dynamics occur beyond predetermined biological paradigms. Both auto- and sympoiesis are equally criticized by Ettinger for their fundamental absence of ethics and the lack of trans-in-spirational sensibilities such as trust, openness, fragilization, carriance, resonance, and subreal awareness. Matrixial breath is *beyond balance* but consists of a fundamental *asymmetry* – in welcoming the other and the unknowingness of what effects the other's breath will have on the closed territory of the subject and to which (other) territorial spaces it can lead to. Wilhelm Reich has aptly stated that "as long as people fear the streaming of living Life in their bodies, they will fear truth and avoid it by all means" (Reich 1953, 192). As long as people proclaim a life "in harmony" and are not prepared to open the closed boundaries of their identity to the indeterminate breath of the other, true humanity will not be realized. Ettingerian environmental ethics are thus opposed to a planetary holism that views the earth as an interconnected ecosystem in which all participants play the same role to maintain the ideal state of ecological equilibrium. Matrixial breath turns precisely against any homeostatic ideal and thermodynamic equilibrium. In this context, Ettinger speaks of unsettling any equilibrium through a co-poietic de-territorialization with and through the other: "In the *matrix*, libidinal investment is directed towards co-emergence; towards a continuum of creation and *disruption in equilibrium*; towards the tuning of processes of separation without rejection and closeness without fusion" (Ettinger 1994, 50). Ettingerian environmental ethics would refer to the "whole" planetary scale of its considerations only to illustrate the multiplicity of others that the I encounters as (non-)Is: the existence of the face, the transcendence of God, and the immanence of planetary matter. Trans-in-spiration also does not consider itself an "interconnection," similar to the systemic relationality in Heidegger's fourfold. Floating-into-each-other is beyond any technological link: the planet is not a machine, the other is not a switching point. The other always remains the other – unknown, encountering me, carrying me, then disappearing again. There is no homeostatic stability here, but rather a reality that Gilbert Simondon has described as "metastability":

> Neither mechanism nor energetism, which are theories of identity, can completely account for this reality (…) Within the living itself, there is a more complete regime of *internal resonance*, one that requires permanent communication and that maintains a metastability that is a condition of life
>
> (Simondon 2009, 6–7)

and further: "a regime of metastability is not only maintained by the individual, but *carried* by it" (8). Metastability is not defined by determined identities but by changing, carrying, and overlapping resonances – the procedures of trans-in-spiration. Here, it is important to note that Simondon discovers true aliveness, a "condition of life," in this resonating reality. Therefore, any holism that strives for balance and harmony always remains detached from aliveness; any holism will never truly embrace the environment, perceive it, open up to it, and be ready to be carried by its resonating drifts and currents. While holistic environmental ethics remain stuck in the androcenic ideologies of commonality and quasi-equal "entanglements," any truly humane approach to a responsible relationship between humans and nature can never be conceived as equal but must always start from the asymmetry of the other and begin to float into its her/his/its otherness.

*

At this point, a link between Ettingerian environmental ethics and the notion of perichoresis in feminist theology can be drawn. Perichoresis (composed of the Greek περί – "around" – and χωρέω – "go forward, travel") originally refers to the relationship of the three Trinitarian persons and characterizes this relationship as "a cyclical movement, a revolving action such as the revolution of a wheel" (Johnson 2018, 231). This happens in contrast to the prevailing male-patriarchal hierarchy of Father, Son, and Holy Spirit. With perichoresis, an "interweaving of things with each other" takes place, a "pervading exchange of life," wherein "each person encompasses the others, is coinherent with the others in a joyous movement of shared life" (231). Notably, perichoresis does not promote an exclusive view of the divine Trinity but rather a *field of relationality* open to God, human beings, and the world. Tim Ingold follows a similar path when he describes perichoresis as "referring to the dance in which each of the divine persons of the Trinity revolves around the others, generating in their circulation a kind of *longing* that springs not from any one person in particular but from their mutual *adoration*" (Ingold 2015, 147). This reveals a relationship in which entities neither confront nor merge holistically but partake in a fragile dance-like interplay in which each of the three divine persons flows into and around each other, attracting and carrying away each other again.[1] The identities of Father, Son, and Holy Spirit can never be clearly defined; they always form themselves anew and figure themselves differently: "Interpreted perichoretically, the trinitarian persons form their own unity by themselves in the circulation of the divine life" (Moltmann 1993, 175). He further elaborates:

> If the concept of person comes to be understood in trinitarian terms – that is, in terms of relation and historically – then the Persons do not only subsist in the common divine substance; they also exist in their relations to the other Persons. More – they are alive in one another and

> through the others in each several case. This idea found expression in the early church's doctrine about the *immanentia* and *inexistentia* of the trinitarian Persons: *intima et perfecta inhabitatio unius personae in alia.*
>
> (174)[2]

Ettinger also argues perichoretically when she states that in matrixial "joining-in-differentiating (…) a spiritual knowledge of the Other and the Cosmos is born and revealed" (Ettinger 2005, 708). Matrixial forms of thought and action always occur through a relational interplay, forming intensities, transitions, and free spaces beyond predetermined identity positions or holistic uniformity. Ettingerian environmental ethics that occurs in trans-in-spiration with the other can thus also be understood as a consistently relational and response-able practice initiated by opening for and carrying the other's face, then touched by the transcendence of God, and placed in the expanses of the earth. In the trans-in-spirational process, the other-as-(non-)I is welcomed through the ethical organ of the lung. Existing territorializations are transformed, and new identities begin to unfold: more-than-human particles float with human, more-than-human, and unknown particles to form *clouds of identity* and *constellations of corporeality*: matrixial condensations in the loving medium of air.

*

Most fundamentally, Ettingerian environmental ethics aim to establish the moment of encounter with the other as an anchor point for new ethical thought and action. Not retreat, but embracement. Such ethics is formed in realizing that encountering the other does not necessitate death or sacrifice as the other is never a "total stranger": the other is always only "almost strange," a (non-)I, and thus a particular part of my self. Ettingerian environmental ethics takes the first step of unclosing for the reception of the other: becoming fragile, permeable, and porous. Territorial boundaries begin to dissolve; a shared aerial dimension opens up in which the other is carried by me (just as I am carried by the other): a floating-into-each-other takes place. *Being wafted* by the other, which initiates all further trans-in-spirational movements, marks a truly perichoretic procedure: a dwelling-of-the-other-in-the-self via approaching and passing each other. In such processes, not only the other humans but also the other living beings, the transcendence of God, and the inanimate matter waft at me and manifest themselves as a distant yet proximal part of me. This does not mean a holistic appropriation of the more-than-human world or pantheistic romanticism. Instead, it manifests that the human is a carrier and can only exist through welcoming and absorbing the other. In such a receptive "jointness-in-difference," anesthesia and androcenic bliss cease and give way for a breathing space in which the human being recognizes her-/himself in the other and takes responsibility for this other: the other is always already much closer than conveyed by the hegemonic principles of philosophy and the natural sciences. Ettingerian environmental ethics thus present an alternative to a holism full of "empty empathy" or "empathy-without-compassion" (Ettinger 2016b, 287). In an ethical approach that dares to breathe in a traumatic vacuum and open itself to the ambiguity of the other, an "empathy within compassion" (118) unfolds, which is about immediate humanity: being affected by the other, by her/his/its wounds, her/his/its languages, her/his/its expressions, and her/his/its existence. A reception of the other that is made possible *only in this moment* of androcenic barbary, which initiates anesthetic intubation. To realize Ettingerian environmental ethics, only one step is necessary: to make the decision, this time and "despite everything," not to use a breathing tube. Let me fully breathe, perceive, and welcome the other! Humanity lies beyond homeostatic indifference.

Notes

1 For a comprehensive introduction to perichoresis and modernity, see Moltmann 1997, 111–24.
2 Translated as: "The intimate and perfect dwelling of one person in another."

References

Anderlini-D'Onofrio, Serena. 2009. *Gaia and the New Politice of Love: Notes for a Poly Planet*. Berkeley, CA: North Atlantic Books.

Anthropocene Working Group. 2022. "Crawford Lake, Canada." Accessed 26 May 2024. https://www.anthropocene-curriculum.org/the-geological-anthropocene/site/crawford-lake.

Apollonio, Umbro, ed. 2001. *Futurist Manifestos*. New York, NY: Distributed Art Publishers.

Arendt, Hannah. 2006. *Eichmann in Jerusalem: A Report on the Banality of Evil*. London: Penguin Books.

Aristotle. 1952. *Meteorologica*. Translated by H. D. P. Lee. Cambridge, MA: Harvard University Press.

Ausländer, Rose. 1987. *Im Atemhaus Wohnen: Gedichte*. Frankfurt am Main: Fischer.

Bacon, Francis. 1855. *Novum Organum: Sive Indicia Vera de Interpretatione Naturae*. Oxford: Typographeo Academico.

Barthes, Roland. 1977. *Image, Music, Text: Essays Selected and Translated by Stephen Heath*. London: Fontana Press.

Barthes, Roland. 1991. "Cy Twombly: Works on Paper." In *The Responsibility of Forms: Critical Essays on Music, Art, and Representation*, 157–76. Berkeley, CA and Los Angeles, CA: University of California Press.

Benjamin, Walter. 2007. *Illuminations: Essays and Reflections*. Translated by Harry Zohn, edited and with and introduction by Hannah Arendt. New York, NY: Schocken Books.

Bogue, Ronald. 2003. *Deleuze on Music, Painting and the Arts*. New York, NY: Routledge.

Böttigheimer, Christoph. 2006. "Der Verantwortungsbegriff Levinas' und der Stellvertretertod Jesu." In *Für das Unsichtbare sterben: Zum 100. Geburtstag von Emmanuel Levinas*, edited by Norbert Fischer, 43–59. Paderborn: Ferdinand Schöningh.

Boyle, Robert. 1660. *New Experiments Physico-Mechanicall: Touching the Spring of the Air, and Its Effects*, Oxford: H. Hall.

Braidotti, Rosi, and Maria Hlavajova. 2018. *Posthuman Glossary*. London and New York, NY: Bloomsbury Academic.

Buber, Martin. 1970. *I and Thou*. Translated by Walter Kaufmann. New York, NY: Charles Scribner's Sons.

Caraco, Albert. 1982. *Le Bréviaire du Chaos*. Paris: L'Age d'Homme.

Cassin, Barbara. 2014. *Dictionary of Untranslatables: A Philosophical Lexicon*. Translated by Steven Rendall, Christian Hubert, Jeffrey Mehlman, Nathanael Stein, and Michael Syrotinski. Princeton, NJ: Princeton University Press.

Celan, Paul. 1952. *Mohn und Gedächtnis: Gedichte.* Stuttgart: Deutsche Verlags-Anstalt.
Celan, Paul. 1964. *Von Schwelle zu Schwelle: Gedichte.* Stuttgart: Deutsche Verlags-Anstalt.
Celan, Paul. 1995. *Breathturn.* Translated by Pierre Joris. Los Angeles, CA: Sun & Moon Press.
Celan, Paul. 1996. *Selected Poems.* Translated by Michael Hamburger. London: Penguin Books.
Celan, Paul. 2001. *Selected Poems and Prose of Paul Celan.* Translated by John Felstiner. New York, NY and London: W.W. Norton.
Cioran, E. M. 1968. *The Temptation to Exist.* Translated by Richard Howard. Chicago, IL: Quadrangle Books.
Cioran, E. M. 1986. *The Trouble With Being Born.* Translated by Richard Howard. New York, NY: Seaver Books.
Cioran, E. M. 1987. *History and Utopia.* Translated by Richard Howard. New York, NY: Arcade Publishing.
Cioran, E. M. 1992. *On the Heights of Despair.* Translated by Ilinca Zarifopol-Johnston. Chicago, IL and London: University of Chicago Press.
Cioran, E. M. 2010. *A Short History of Decay.* Translated by Richard Howard. London: Penguin Classics.
Clark, Timothy. 1997. *The Theory of Inspiration: Composition as a Crisis of Subjectivity in Romantic and Post-Romantic Writing.* Manchester: Manchester University Press.
Crockett, Clayton. 2018. *Derrida after the End of Writing: Political Theology and New Materialism.* New York, NY: Fordham University Press.
Crutzen, Paul J., and Eugene F. Stoermer. 2000. "The 'Anthropocene'." *IGBP Newsletter* 41: 17–18.
Deleuze, Gilles, and Félix Guattari. 1987. *A Thousand Plateaus: Capitalism and Schizophrenia.* Translated by Brian Massumi. Minneapolis, MN and London: University of Minnesota Press.
Deleuze, Gilles, and Felix Guattari. 2000. *Anti-Oedipus: Capitalism and Schizophrenia.* Translated by Robert Hurley, Mark Seem, and Helen R. Lane. Minneapolis, MN: University of Minnesota Press.
Deleuze, Gilles. 1992. "Postscript on the Societies of Control." *JSTOR* 59: 3–7.
Deleuze, Gilles. 1994. *Difference and Repetition.* Translated by Paul Patton. New York, NY: Columbia University Press.
Deming, Barbara. 1984. *We Are All Part of One Another: A Barbara Deming Reader.* Edited by Jane Meyerding. Philadelphia, PA: New Society Publishers.
Dierks, Jan, Konrad Ott, and Lieske Voget-Kleschin. 2016. *Handbuch Umweltethik.* Stuttgart: J. B. Metzler.
Dostoevsky, Fyodor. 1960. *Notes From Underground.* Translated by Ralph E. Matlaw. New York, NY: Dutton.
Edwards, Paul N. 2010. *A Vast Machine: Computer Models, Climate Data, and the Politics of Global Warming.* Cambridge, MA and London: MIT Press.
Ettinger, Bracha L. 1992. "Matrix and Metramorphosis." *Differences: Journal of Feminist Cultural Studies* 4 (3): 176–208.
Ettinger, Bracha L. 1993. *Time Is the Breath of the Spirit: Emmanuel Levinas in conversation with Bracha Lichtenberg Ettinger.* Oxford: Museum of Modern Art.
Ettinger, Bracha L. 1994. "The Becoming Threshold of Matrixial Borderlines." In *Travellers' Tales: Narratives of Home and Displacement*, edited by George Robertson, Melinda Mash, Lisa Tickner, Jon Bird, Barry Curtis, and Tim Putnam, 36–60. London: Routledge.

Ettinger, Bracha L. 1999a. "Trauma and Beauty: Trans-Subjectivity in Art." *n.paradoxa* 3: 15–23.
Ettinger, Bracha L. 1999b. "Traumatic Wit(h)ness-Thing and Matrixial Co/in-habit(u)ating." *Parallax* 5 (1): 89–98.
Ettinger, Bracha L. 2001a. "Working-Through. A Conversation Between Craigie Horsfiel and Bracha Lichtenberg Ettinger." *The Drawing Center's Drawing Papers* 24: 37–62.
Ettinger, Bracha L. 2001b. "Matrixial Gaze and Screen: Other than Phallic and Beyond the Late Lacan." In Bodies of Resistance, edited by Laura Doyle, 103–143. Evanston, IL: Northwestern University Press.
Ettinger, Bracha L. 2005. "Copoiesis." *Ephemera: Theory & Politics in Organization* 5 (x): 703–13.
Ettinger, Bracha L. 2006a. "The *Heimlich*." In *The Matrixial Borderspace,* edited by Brian Massumi, 157–61. Minneapolis, MN: University of Minnesota Press.
Ettinger, Bracha L. 2006b. "Weaving a Woman Artist with-in the Matrixial Encounter-Event." In *The Matrixial Borderspace,* edited by Brian Massumi, 173–200. Minneapolis, MN: University of Minnesota Press.
Ettinger, Bracha L. 2006c. "From Proto-Ethical Compassion to Responsibility: Besidedness and the Three *Primal* Mother-Phantasies of Not-Enoughness, Devouring and Abandonment." *Athena: Filosofijos Studijos* 2: 100–36.
Ettinger, Bracha L. 2010. "Transgressing With-In-To the Feminine." In *Feminist Readings of Antigone*, edited by Fanny Söderbäck, 195–214. Albany, NY: State University of New York Press.
Ettinger, Bracha L. 2011. "Uncanny Awe, Uncanny Compassion and Matrixial Transjectivity beyond Uncanny Anxiety." *French Literature Series* 38: 1–30.
Ettinger, Bracha L. 2014. "Demeter-Persephone Complex, Entangled Aerials of the Psyche, and Sylvia Plath." *ESC: English Studies in Canada* 40 (1): 123–54.
Ettinger, Bracha L. 2015a. "Carriance, Copoiesis and the Subreal." In *And My Heart-Wound Space*, edited by Bracha L. Ettinger, 343–51. Leeds: The Wild Pansy Press.
Ettinger, Bracha L. 2015b. "And My Heart, Wound-Space With-In Me: The Space of Carriance." In *And My Heart-Wound Space*, edited by Bracha L. Ettinger, 353–67. Leeds: The Wild Pansy Press.
Ettinger, Bracha L. 2016a. "Art as the Transport-Station of Trauma." In *Interdisciplinary Handbook of Trauma and Culture*, edited by Yochai Ataria, David Gurevith, and Haviva Pedaya, 151–60. Cham: Springer.
Ettinger, Bracha L. 2016b. "Laius Complex and Shocks of Maternality: With Franz Kafka and Sylvia Plath." In *Interdisciplinary Handbook of Trauma and Culture*, edited by Yochai Ataria, David Gurevith, and Haviva Pedaya, 279–302. Cham: Springer.
Ettinger, Bracha L. 2016c. "Art in a Time of Atrocity." *The New York Times*, 16 December. Accessed 26 May 2024. https://www.nytimes.com/2016/12/16/opinion/art-in-a-time-of-atrocity.html.
Ettinger, Bracha L. 2018. "The Co/Sympoietic Human." Lecture presented at the Lecture Series *Human After Man*, Academy of Fine Arts, Munich, 6 November.
Ettinger, Bracha L. 2019. "Beyond the Death-Drive, Beyond the Live-Drive – Being-Toward-Birthing with Being-Toward-Birth: Copoiesis and the Matrixial Eros – Metafeminist Notes." In *Aberrant Nuptials: Deleuze and Artistic Research*, edited by Paolo de Assis and Paolo Giudici, 183–214. Leuven: Leuven University Press.
Ettinger, Bracha L. 2020. *Matrixial Subjectivity, Aesthetics, Ethics: Volume 1 1990–2000*, edited by Griselda Pollock. London: Palgrave Macmillan.
Fardy, Jonathan. 2020. *Laruelle and Art: The Aesthetics of Non-Philosophy*. London and New York, NY: Bloomsbury Academic.

Foucault, Michel. 1995. *Discipline and Punish: The Birth of the Prison*. Translated by Alan Sheridan. New York, NY: Vintage Books.
Freud, Sigmund. 1955. "Beyond the Pleasure Principle (1920)." In *The Standard Edition of the Complete Psychological Works of Sigmund Freud*, translated by James Strachey, Volume XVIII, 3–66. London: Hogarth Press.
Freud, Sigmund. 1958. "Formulations on The Two Principles of Mental Functioning (1911)." In *The Standard Edition of the Complete Psychological Works of Sigmund Freud*, translated by James Strachey, Volume XII, 213–26. London: Hogarth Press.
Freud, Sigmund. 1959. "Civilized Sexual Morality and Modern Nervous Illness (1908)." In *The Standard Edition of the Complete Psychological Works of Sigmund Freud*, translated by James Strachey, Volume IX, 177–204. London: Hogarth Press.
Freud, Sigmund. 1961. "Civilization and Its Discontents (1929 [1930])." In *The Standard Edition of the Complete Psychological Works of Sigmund Freud*, translated by James Strachey, Volume XXI, 59–148. London: Hogarth Press.
Freud, Sigmund. 1964. "New Introductory Lectures on Psycho-Analysis: XXXIII Femininity (1933 [1932])." In *The Standard Edition of the Complete Psychological Works of Sigmund Freud*, translated by James Strachey, Volume XXII, 112–35. London: Hogarth Press.
Glazebrook, Trish. 2016. "Gynocentric Bio-Logics: Anthropocenic Abjectification and Alternative Knowledge Traditions." *Telos* 177: 61–82.
Graessle, William. 2018. "Lynn White Jr.'s Medieval Heavy Plow: An Instrument of Agricultural Innovation, Population Growth, and Urbanization in High Medieval Western Europe." *A Journal of History* 2: 27–59.
Guattari, Félix. 1973. "La place du sigifiant dans l'institution." *Lettres de l'École freudienne de Paris* (Congrès de la Grande Motte 1 au 4 novembre 1973). Accessed 26 May 2024. http://ecole-lacanienne.net/wp-content/uploads/2016/04/22_n-17-groupe-de-travail-sur-l-institution-du-lundi-5-11-am.pdf.
Guattari, Félix. 1995. *Chaosmosis: An Ethico-Aesthetic Paradigm*. Translated by Paul Bains and Julian Pefanis. Bloomington, IN and Indianapolis, IN: Indiana University Press.
Han, Byung-Chul. 2014. "'Tut mir leid, aber das sind Tatsachen'." *Die Zeit*. 7 September. Accessed 26 May 2024. https://www.zeit.de/zeit-wissen/2014/05/byung-chul-han-philosophie-neoliberalismus?utm_referrer=https%3A%2F%2Fwww.google.com%2F.
Han, Byung-Chul. 2021. *The Palliative Society: Pain Today*. Translated by Daniel Steuer. Cambridge and Medford, MA: Polity Press.
Haraway, Donna J. 2016. *Staying with the Trouble: Making Kin in the Chthulucene*. Durham, NC and London: Duke University Press.
Harman, Graham. 2015. "Art and OOObjecthood: Graham Harman in Conversation with Christoph Cox and Jenny Jaskey." In *Realism Materialism Art*, edited by Christoph Cox, Jenny Jaskey, and Suhail Malik, 97–122. Berlin: Sternberg Press.
Harman, Graham. 2017. "For a Thought of Objects: Interview with Graham Harman." *Stream* 4: 29–42.
Heidegger, Martin. 1973. "Art and Space: Translated by Charles H. Seibert." *Continental Philosophy Review* 6: 3–8.
Heidegger, Martin. 1993a. "The Origin of the Work of Art." In *Basic Writings*, edited by David Farrell Krell, 139–212. New York, NY: Harper Collins Publishers.
Heidegger, Martin. 1993b. *Basic Concepts*. Translated by Gary E. Aylesworth. Bloomington, IN and Indianapolis, IN: Indiana University Press.
Heidegger, Martin. 1996. *Being and Time: A Translation of Sein und Zeit*. Translated by Joan Stambaugh. Albany, NY: State University of New York Press.

Heidegger, Martin. 1998. "What Is Metaphysics?" Translated by David Farrell Krell. In *Pathmarks*, edited by William McNeill, 82–96. Cambridge: Cambridge University Press.

Heidegger, Martin. 2001a. "Building, Dwelling, Thinking." In *Poetry, Language, Thought*, translated by Albert Hofstadter, 141–60. New York, NY: Harper Collins Publishers.

Heidegger, Martin. 2001b. "The Thing." In *Poetry, Language, Thought*, translated by Albert Hofstadter, 161–84. New York, NY: Harper Collins Publishers.

Heidegger, Martin. 2001c. "...Poetically Man Dwells..." In *Poetry, Language, Thought*, translated by Albert Hofstadter, 209–27. New York, NY: Harper Collins Publishers.

Heidegger, Martin. 2012. *Bremen and Freiburg Lectures: Insight into That Which Is and Basic Principles of Thinking*. Translated by Andrew J. Mitchell. Bloomington, IN and Indianapolis, IN: Indiana University Press.

Hopkins, Gerard Manley. 2002. *Gerard Manley Hopkins: The Major Works*. Edited by Catherine Phillips. Oxford: Oxford University Press.

Humboldt, Alexander von. 1806. *Ideen zu einer Physiognomik der Gewächse* [no publisher]. Accessed 24 May 2024. https://haab-digital.klassik-stiftung.de/viewer/image/3493357036/2/.

Husserl, Edmund. 2020. *Husserliana 43. Studien zur Struktur des Bewusstseins, Teilband III. Wille und Handlung*. Edited by Ullrich Melle and and Thomas Vongehr. Dordrecht: Springer.

Ingold, Tim. 2015. *The Life of Lines*. London and New York, NY: Routledge.

Johnson, Elizabeth A. 2018. *She Who Is: The Mystery of God in Feminist Theological Discourse*. New York, NY: Crossroad Publishing Company.

Kaiser, Birgit M., and K. Thiele. 2018. "If You Do Well, Carry! The Difference of the Humane: An Interview with Bracha L. Ettinger." *philoSOPHIA* 8 (1): 101–25.

Kant, Immanuel. 2000. *Critique of Pure Reason*. Translated and edited by Paul Guyer and Allen W. Wood. Cambridge: Cambridge University Press.

Khubchandani, Jagdish, Sushil Sharma, and James H. Price. 2021. "COVID-19 Pandemic and the Burden of Internet Addiction in the United States." *Psychiatry International* 2: 402–9. https://doi.org/10.3390/psychiatryint2040031.

"Kontrollgesellschaft." 2000. *Terz* 6. Accessed 24 May 2024. https://terz.org/2000/06/kontrollgesellschaft.html.

Laruelle, François. 2012. "Is Thinking Democratic? Or, How to Introduce Theory into Democracy." In *Laruelle and Non-Philosophy*, edited by John Mullarkey and Anthony Paul Smith. Edinburgh: Edinburgh University Press.

Laruelle, François. 2013. *Dictionary of Non-Philosophy*. Translated by Taylor Adkins. Minneapolis, MN: Univocal Publishing.

Laruelle, François. 2017. *Principles of Non-Philosophy*. Translated by Nicola Rubczak and Anthony Paul Smith. London and New York, NY: Bloomsbury Publishing.

Laruelle, François. 2018. *A Biography of Ordinary Man: On Authorities and Minorities*. Cambridge: Polity Press.

Levinas, Emmanuel. 1969. *Totality and Infinity*. Translated by Alphonso Lingis. Pittsburgh, PA: Duquesne University Press.

Levinas, Emmanuel. 1991. *Otherwise than Being or Beyond Essence*. Translated by Alphonso Lingis. Dordrecht: Springer.

Levinas, Emmanuel. 1999. *Alterity and Transcendence*. Translated by Michael B. Smith. New York, NY: Columbia University Press.

Lorde, Audre. 1988. *A Burst of Light: Essays by Audre Lorde*. Ithaca, NY: Firebrand Books.

Lu, Dong Wei, Jenn Wu Wang, and Andrew Chih Wei Huang. 2010. "Differentiation of Internet Addiction Risk Level Based on Autonomic Nervous Responses: The Internet-Addiction Hypothesis of Autonomic Activity." *Cyberpsychology, Behavior, and Social Networking* 13 (4): 371–78. https://doi.org/10.1089=cyber.2009.0254.

Lyotard, Jean-François. 1993. *Libidinal Economy*. Translated by lain Hamilton Grant. Bloomington, IN and Indianapolis, IN: Indiana University Press.

Margulis, Lynn. 2001. *The Symbiotic Planet: A New Look at Evolution*. London: Phoenix.

Margulis, Lynn, and Dorion Sagan. 1997. *Microcosmos: Four Billion Years of Evolution from Our Microbial Ancestors*. Berkeley, CA: University of California Press.

Margulis, Lynn, and Dorion Sagan. 2000. *What Is Life?* Berkeley, CA and Los Angeles, CA: University of California Press.

Maturana Humberto R., and Francisco J. Varela. 1980. *Autopoiesis and Cognition: The Realization of the Living*. Dordrecht: D. Reidel Publishing Company.

Maturana Humberto R., and Francisco J. Varela. 1992. *The Tree of Knowledge: The Biological Roots of Human Understanding*. Boston, MA: Shambala.

McCarthy, Francine M. G., Timothy S. Patterson, Martin J. Head, Nicholas L. Riddick, Brian F. Cumming, Paul B. Hamilton, Michael F. J. Pisaric, Cale A. C. Gushulak, Peter R. Leavitt, Krysten M. Lafond, Brendan M. Llew-Williams, Matthew G. Marshall, Autumn Heyde, Paul Michael Pilkington, Joshua Moraal, Joseph I. Boyce, Nawaf A. Nasser, Carling Ruth Walsh, M. Garvie, Sarah L. Roberts, Neil L. Rose, Andrew B. Cundy, Paweł Gaca, Andrew Milton, Irka Hajdas, Carley Crann, Arnoud Boom, Sarah A. Finkelstein, and John H. McAndrews. 2023. "The Varved Succession of Crawford Lake, Milton, Ontario, Canada as a Candidate Global Boundary Stratotype Section and Point for the Anthropocene series." *The Anthropocene Review* 10 (1): 146–76.

Merchant, Carolyn. 1983. *The Death of Nature: Women, Ecology, and the Scientific Revolution*. New York, NY: Harper & Row.

Mitchell, Andrew J. 2015. *The Fourfold: Reading the Late Heidegger*. Evaston, IL: Northwestern University Press.

Moltmann, Jürgen. 1993. *The Trinity and the Kingdom: The Doctrine of God*. Minneapolis, MN: Fortress Press.

Moltmann, Jürgen. 1997. *Gott im Projekt der modernen Welt: Beiträge zur öffentlichen Relevanz der Theologie*. Gütersloh: Chr. Kaiser/Gütersloher Verlagshaus.

Morton, Timothy. 2013. *Hyperobjects: Philosophy and Ecology After the End of the World*. Minneapolis, MN: University of Minnesota Press.

Morton, Timothy. 2016. *Dark Ecology: For a Logic of Future Coexistence*. New York, NY: Columbia University Press.

Musil, Robert. 1995. *The Man Without Qualities*. Translated by Sophie Wilkins. New York, NY: Vintage Books.

Nieuwenhuis, Marijn. 2015. "On One Breath All Depend." *Journal of Narrative Politics* 1 (2): 167–79.

Nieuwenhuis, Marijn. 2018. "Atmospheric Governance: Gassing as Law for the Protection and Killing of Life." *Environment and Planning D: Society and Space* 36 (1): 78–95.

Ọláléyẹ, Désọlá. 2023. "'Babygirling' and the Pitfalls of the Soft Life Brigade." *The Elephant – Explaining Society to the People*. Accessed 24 May 2024. https://www.theelephant.info/culture/2023/05/15/babygirling-and-the-pitfalls-of-the-soft-life-brigade/.

Pollock, Griselda. 2006. "Femininity: Aporia or Sexual Difference." In *The Matrixial Borderspace*, edited by Brian Massumi, 1–40. Minneapolis, MN: University of Minnesota Press.

Prior, Moody E. 1954. "Bacon's Man of Science." *Journal of the History of Ideas* 15 (3): 348–70.

Reder, Michael, Andreas Gösele, Lukas Köhler, and Johannes Wallacher. 2019. *Umweltethik: Eine Einführung in globaler Perspektive*. Stuttgart: W. Kohlhammer.

Reich, Wilhelm. 1953. *The Murder of Christ: Volume One of The Emotional Plague of Mankind*. New York, NY: Simon & Schuster.

Roszak, Theodore. 2001. *The Voice of the Earth: An Exploration of Ecopsychology*. Grand Rapids, MI: Phanes Press.

Ruyer, Raymond, and R. Scott Walker. 1980. "The Status of the Future and the Invisible World." *Diogenes* 28 (109): 37–53.

Scheler, Max. 1976. *Späte Schriften*. Edited by Manfred S. Frings. Bern: Francke.

Severson, Eric R. 2011. "Deesis: Levinas, Language and Psychology." Unpublished paper.

Shafak, Elif. 2021. "Apathy Biggest Threat to Humanity, Says Elif Shafak." *Times of India*, 19 September. Accessed 24 May 2024. http://timesofindia.indiatimes.com/articleshow/86332065.cms?utm_source=contentofinterest&utm_medium=text&utm_campaign=cppst.

Simondon, Gilbert. 2009. "The Position of the Problem of Ontogenesis. Translated by Gregory Flanders." *Parrhesia* 7: 4–16.

Škof, Lenart. 2015. *Breath of Proximity: Intersubjectivity, Ethics and Peace*. Dordrecht: Springer.

Sloterdijk, Peter. 2009a. "Talking to Myself About the Poetics of Space." *Harvard Design Magazine* 30: 1–8.

Sloterdijk, Peter. 2009b. *Terror from the Air*. Translated by Amy Patton and Steve Corcoran. Los Angeles, CA: Semiotext(e).

Sloterdijk, Peter. 2013. *In the World Interior of Capital: For a Philosophical Theory of Globalization*. Translated by Wieland Hoban. Cambridge: Polity Press.

Spahr, Juliana. 2005. *This Connection of Everyone with Lungs*. Berkeley, CA and Los Angeles, CA: University of California Press.

Tarr, Béla. 2012. *The Turin Horse*. London: Artificial Eye.

Tarkovsky, Andrei. 2002. *Stalker*. London: Artificial Eye.

Teutsch, Gotthard M. 1985. *Lexikon der Umweltethik*. Göttingen: Vandenhoeck und Ruprecht; Düsseldorf: Patmos-Verlag.

Tolstoy, Leo. 2010. *War and Peace*. Translated by Louise and Aylmer Maude. Oxford: Oxford University Press.

Virilio, Paul. 1994. *Bunker Archeology*. Translated by George Collins. New York, NY: Princeton Architectural Press.

Von Redecker, Eva. 2022. "Das Wagnis des Weichbleibens: Kevin Rittberger im Gespräch mit Eva von Redecker." *Schauspiel Hannover Magazin 10*. Accessed 24 May 2024. https://staatstheater-hannover.de/de_DE/das-wagnis-des-wachbleibens.

Welker, Michael. 1994. *God the Spirit*. Translated by John F. Hoffmeyer. Minneapolis, MN: Fortress Press.

White, Lynn Jr. 1967. "The Historical Roots of Our Ecologic Crisis." *Science* 155 (3767): 1203–7.

Zhong, Raymond. 2024. "Are We in the 'Anthropocene,' the Human Age? Nope, Scientists Say." *New York Times*, 5 March 2024, Accessed 20 April 2025. https://www.nytimes.com/2024/03/05/climate/anthropocene-epoch-vote-rejected.html.

Žižek, Slavoj. 2007. "Censorship Today: Violence, or Ecology as a New Opium for the Masses." *Lacan.com.* Accessed 24 May 2024. https://www.lacan.com/zizecology1.htm.

Žižek, Slavoj. 2018. *Like a Thief in Broad Daylight: Power in the Era of Post-Human Capitalism.* London: Penguin.

Zupančič, Alenka. 2018. "The Apocalypse Is (Still) Disappointing." *S: Journal of the Circle for Lacanian Ideology Critique* 11: 16–30.

Zupančič, Alenka. 2019. "On Sex Without Identity: Feminist Politics and Sexual Difference". *Los Angeles Review of Books*, 7 October. Accessed 24 May 2024. https://lareviewofbooks.org/article/sex-without-identity-feminist-politics-sexual-difference/.

Index

FSC
www.fsc.org
MIX
Papier | Fördert
gute Waldnutzung
FSC® C083411

Kolibri 360 GmbH
Ferdinand-Jühlke-Straße 7
99095 Erfurt, Deutschland
produktsicherheit@kolibri360.de